Chance and Change

CHANCE
and CHANGE
ECOLOGY FOR CONSERVATIONISTS

WILLIAM HOLLAND DRURY JR.

EDITED BY JOHN G. T. ANDERSON
WITH A FOREWORD BY ERNST MAYR

UNIVERSITY OF CALIFORNIA PRESS
Berkeley   Los Angeles   London

University of California Press
Berkeley and Los Angeles, California

University of California Press, Ltd.
London, England

Library of Congress Cataloging-in-Publication Data

Drury, W. H. (William H.)
    Chance and change: ecology for conservationists / William Holland
    Drury Jr. ; edited by John G. T. Anderson ; with a foreword by Ernst
    Mayr.
        p. cm.
    Includes bibliographic references and index.
    ISBN 0–520-21155-3 (cloth : alk. paper)
    1. Ecology.   2. Evolution (Biology).   3. Natural selection.
    I. Anderson, John G. T., 1957– .   II. Title.
    QH541.D77   1998
    577—dc 21   97-14859

Printed in the United States of America
9 8 7 6 5 4 3 2 1

For Mary

Contents

Illustrations

Foreword

William H. Drury Jr. had an extraordinary breadth of knowledge and interest. Even though we normally think of him as an ornithologist, he was actually, by training, a botanist, and his Ph.D. thesis dealt with botanical and geological themes. This triple expertise in botany, geology, and zoology gave him a balanced viewpoint that only a very few ecologists possess.

It is true, not only for scientists but for just about everybody, that one can divide people into observers and nonobservers. Some people are just born observers who see things—exciting, interesting things—that no one else sees. Darwin was one of these gifted observers, and so was Konrad Lorenz. Actually, most naturalists have this ability to some extent, and Bill Drury certainly did. He was not simply a describer who conscientiously recorded all he saw, but whenever he looked at the natural scene he saw puzzling irregularities, unanswered questions, contradictions, and what others call exceptions. And this made him question the ruling philosophy of science.

In the 1940s and 1950s when Bill started his career, biology was dominated by physicalism. In order to be scientific, theories had to be deterministic. They had to be expressed in terms of mathematical formulae. Light, water, temperature, and other physical factors were the determinants of ecology, and everything obeyed universal laws. As someone has said, biology at that period was suffering from physics-envy.

Bill was far too good a field man and naturalist to accept these dogmas. I am impressed by Bill's willingness to accept pluralistic solutions, to accept the widespread occurrence of stochastic processes, to accept the importance of *qualitative* aspects in addition to the quantitative ones, and to rejoice in unexpected findings and so-called exceptions. I think these approaches and attitudes describe best the special way in which Bill addressed ecological problems.

In holding these views, he was in direct conflict with the Chicago School of Allee, Emerson, Schmidt, and the Parks, who stressed so much the importance of physical factors in controlling populations and distributions, but he was even more in conflict with the highly typological thinking of the Clements-Shelford school of community ecologists. The whole biome concept made little sense to Bill. He couldn't see that it made much difference to a spruce-dominated landscape if you took the moose out of it. For him there was no superorganism called the spruce-moose biome.

If you adopt physicalist thinking, you look for commonalities and universal laws. Yet, the field naturalist is forever impressed by the uniqueness of everything. Anyone who thinks that by understanding the common tern he or she therefore also understands the Arctic tern, the roseate tern, or the least tern is sadly mistaken. And not only does every species have its unique aspects, very often so does a localized population. Saying this, I am not denying the merits of the search for basic principles or generalizations. Yet, one must be aware at all times that predictions in ecology and field natural history are probabilistic. Most of them have far too many exceptions to justify calling such regularities laws. In fact, it is up to the naturalist to determine which kind of processes lead to reasonable predictive outcomes and which others do not.

Bill was a superb student of behavior. I remember when he was working on the killdeer; every time I saw him he told me of some interesting new behavior that he had discovered. I regret that he never published this work in full. I believe that it was his thorough understanding of behavior that made him such a sound ecologist. The importance of behavior is now being realized by a number of ecologists, and there are now several texts in this field titled "Behavioral Ecology." But Bill surely was one of the first to discover the importance of behavior to ecology.

Bill became an ecologist at an exciting time. It was the time of the beginning of the great controversies in the field, most of which, it seems to me, are still not settled. Even though in his own work Bill could not make a contribution to all of these problems, he was certainly *interested* in all of them.

With his background in botany, he was particularly interested in the controversy between the followers of Clements and Gleason. For Clements, a plant community was a superorganism. According to the well-known claim of the wholeness of the biome, you couldn't take a species out of a biome without upsetting the whole biome. The followers of Gleason, by contrast, contended that each species simply fills its own niche, and the coexistence of certain species in communities is simply the result of the similarity of their niche requirements.

The physicalists believed in the inevitability of certain sequences of succession, and as determinists, they were convinced that all successions would end in a well-defined climax community. This never made sense to Bill, and of course, the typological thinking of the Clements school has since been thoroughly refuted in many modern analyses. Biomes and climax are strictly typological concepts, which neither allow gradations into other kinds of communities nor recognize the enormous amount of variation and change. Bill saw this from the very beginning, thereby avoiding the determinism and typology of the prevailing schools.

In the recent history of ecology, mathematical formulae and experiments under rather specialized laboratory conditions have often provided us with new insights and have added to the rigor of research. However, the book and the laboratory must be checked at all times against the real world. And this is where Bill made his greatest contributions. Years of intensive fieldwork had taught him what goes on in nature, which processes obey certain regularities and which do not. His superb gift for observation told him which textbook theorems he could trust and which he shouldn't.

Bill finally decided to make his insights available to others and began writing this book. His novel approach was to correlate landscape evolution with ecological principles. His diversified training in ornithology, botany, geology, and geography made him uniquely qualified to take this approach. He put all his ideas on paper, some in quite finished form, others evidently still in need of revision. It was at this point that illness struck him and prevented further work. The manuscript was, however, so far advanced that it would have been a real loss to discard it. So his wife, Mary, obtained my help and that of his friends at the College of the Atlantic to edit the manuscript. It was John Anderson, in particular, who finally cast it into publishable form. Some new sections had to be written to fill gaps, peripheral material had to be eliminated to provide better cohesion and continuity, and, indeed, a great deal of rewriting and other editorial work had to be done by Anderson. All of Bill's friends are deeply grateful to Anderson for making the publication of Bill's thoughts possible. There is no doubt that this book will have a wonderfully stimulating impact on ecology.

Ernst Mayr
1997

Editor's Introduction

Chance and Change is the result of a lifetime in the field and in the classroom —and for Bill Drury, there was often very little difference between the two. Bill firmly believed that a real understanding of the out-of-doors could only come from spending as much time there as possible. He felt that an excessive concentration on learning by rote and an emphasis on abstractions both confused students in general and turned many away from an appreciation and understanding of science. In the three years that I was privileged to team-teach with him, I was continually fascinated by his ability to reach avowed "antiscience" students. He would take them out bird-watching, and, as the morning wore on, we would talk about geology, and botany, and the ways that tides and currents worked the shore. Gradually the students would find themselves doing and enjoying science, without ever being sure just how it had happened.

Bill taught by example and riddle. He often commented that "nature has enough variability that you can find evidence for anything if you look hard enough." Hasty observations and excessive reliance on theory would inevitably lead to false conclusions. Bill strongly advocated long-term studies at multiple locations. His work in New England included an extensive review of more than one hundred years of natural history data on gulls, combined with his own twenty years' worth of field surveys. At the time of his death he was planning a monitoring program extending into the next century that would examine trends in songbird populations.

Many who are familiar with Bill's ornithological work will be surprised at the relatively few references to seabird ecology. Conversations with his colleagues and students make it clear that most expected that "the book" would be "Bill's Gull Book." In fact there was very little about gulls among the manuscript pages, and much of what was there consisted of examinations of other people's work. I have attempted to retain everything relating to seabirds that grew directly out of Bill's experience, but it is important to realize that Bill was a botanist and a geologist long before he had the opportunity to focus professionally on birds. The birds are always there, flitting among the trees, passing over the tundra, feeding in the salt marshes; Bill never seems to have had much time for the limitations that standard academic divisions often put on what should interest the observer. To the true naturalist ecology is all-encompassing, and one can often learn more about birds by watching the tide than by simply concentrating on a particular species or population to the exclusion of all else.

Chance and Change seems to have had several beginnings and taken a number of forms during the course of its writing. Bill often tried out selected excerpts on groups of students, and as the students changed, so did the book. Among the manuscript files I found more than half a dozen different hypothetical chapter lists, some with as many as twenty-five chapters proposed. Some of these chapters were almost completed at the time of Bill's death, others were barely in outline form. Essentially what I have attempted to do is retain as many of Bill's ideas as possible without resorting to filling in holes with what I guessed Bill might have said. The result is inevitably only a partial view of the edifice that Bill hoped to construct, but at least it is almost entirely in his own words.

Bill had always intended to illustrate the book himself, but he left the figures until last and unfortunately never got around to them. This placed me in a cruel dilemma. On the one hand some parts of the text needed specific figures to illustrate particular points; on the other hand, Bill had created many wonderful sketches and drawings that I felt should be shared with his readers, but none of these were specifically intended for the text. As a compromise, Cynthia Borden-Chisholm, a former student of Bill's, agreed to do some specific text-related figures. Bill's wife, Mary, searched hundreds of Bill's drawings and slides for pictures that would give the reader a sense of particular species and landscapes mentioned in a given chapter.

·   ·   ·

The book was written for two particular audiences. The first is the group of serious amateur naturalists who wish to go beyond the traditional observational role of the natural historian, and who find much of the abstract theory presented in textbooks at odds with what they see. The second audience is that of professional conservationists who are trying to develop policies and establish alternatives in the face of enormous pressure from conflicting human demands. The book is not intended to be a field guide or a replacement for going outside and seeing things as they are. Rather it is a series of verbal paintings designed to illustrate the causes and consequences of the regularity and disorder that confront us as soon as we step off the paved road or move into the woods.

Ultimately this book is a meditation on human perceptions of the world. Bill was fascinated with our ability to convince ourselves of what "ought" to be—even in the face of vast bodies of "exceptions." For most of his professional career he was in conflict with the ecological establishment. He disagreed strongly with many of the most dearly held beliefs of the founders of modern ecology, and it is ironic that only at the end of his life did an increasing wave of theoreticians begin moving in the direction that years of field experience had pushed Bill long before. Some of the ideas in this book will annoy professional ecologists, who may feel justified in accusing Bill of creating straw men easily savaged. Others may say that Bill is still fighting an old battle that was ending long before his death. To those who feel this way, I gently suggest that you might look at the ecology section of most high school texts and then consider that it is these texts, rather than the contents of seminars, that are the primary sources of academic information for most North Americans. Debates may continue long after the original words have lost their meanings.

Bill was a passionate believer in natural selection as proposed by Darwin and Wallace, and this is the main theme of *Chance and Change*. Just as Dobzhansky said that "nothing in biology makes sense except in the light of evolution," Bill felt that everything in biology made sense in the light of natural selection. He always stressed the disorder and "untidiness" of biology, especially when it was compared to the seeming macroscopic order of chemistry, geology, and Newtonian physics. The sheer variety of environmental conditions in both space and time has led to selection for organisms capable of producing a vast excess of variable young. The tight couplings favored by coevolutionists are there, but they are likely to be the exception rather than the rule. Evolution (which he would sometimes spell "Evilution" in fun) existed as a tendency toward a greater and greater liberation from environmental constraints, including other organisms, rather than the ever-stronger dependencies proposed by community ecologists.

Bill felt that the seeming orderliness of a landscape seen from a distance was usually imposed by the interaction of geological order and particular species characteristics, and that this order tended to break down as soon as one looked closely at any particular location. A deeply embedded human tendency to favor the imposition of structure further complicates any picture. We look out our windows and see forest or prairie and tend to overlook the enormous differences between tree and tree, grass and grass. Natural selection may be crudely referred to as "a reduction in variance," but without this variance the concept has no meaning. Bill saw vital synergism between the variation continually expressed in the vast numbers of organisms and the variety found in any landscape. The vagaries of history—both human and natural—have profound effects on what plants and animals will persist and breed and die in a given place. Some species and some events will pass without notice by human observers; others may transform our perception of a landscape.

Many people feel that a view of life based on natural selection is bleak in the extreme: "Nature, red in tooth and claw." To Bill, natural selection was a source of continuous hope and possibility. Rather than a fragile world balancing on the edge of collapse, Bill saw a vital and dynamic collection of organisms, each with its own strengths and weaknesses, each selected over vast periods of time to do as well as possible under a conflicting array of changes. The lack of many exclusive connections makes most organisms redundant to their fellows. Death, disturbance, and extinction have been the rule, and our fears of "environmental collapse" are probably overstated. Interestingly enough, however, this does not remove our obligation as conservationists. Bill felt that, as the one group of creatures who can clearly, consciously examine alternative futures, we humans have an enormous responsibility to use our abilities to tend and encourage diversity whenever we find it. Rather than waste time arguing about what is "natural" and what is "human," we should get on with the job of protecting and encouraging the delightful array of living things.

Acknowledgments

This book may have one author, but it has taken many midwives to bring it to birth. It is both my pleasure and my responsibility to thank the people who aided Bill during preparation of the bulk of the manuscript and those who assisted me in the task of completing it. Inevitably I will miss some individuals that Bill would have included here as a matter of course, and I apologize for this in advance. Bill exchanged ideas freely with countless individuals, and many parts of the book are developments of those conversations.

Bill's busy teaching schedule was both a source of inspiration and an obstacle to writing, and I know that he would wish to acknowledge the support and encouragement of two longtime friends, Emily V. Wade and Robert G. Goelet, who made it possible for him to take time off to write, and who also supported various stages of his research. Peter Wayne, George Putz, and Dick Henry also provided extensive advice at various stages of the project, and Peter Wayne graciously permitted me to include as part of the final chapter portions of a jointly written but unpublished piece that he and Bill had worked on.

Several chapters of the book were prepared as a direct result of Bill's classes at the College of the Atlantic, and I thank the generations of students who took his courses in ornithology, landforms and vegetation, populations and species, animal behavior, and natural history for posing some of the questions that Bill tries to answer here. Once Bill realized the seriousness of his illness he called together a final band to work with him on particular chapters in an informal seminar. After his death I reassembled the "Book Lunch Bunch" to assist me on the initial job of editing the more than two thousand pages that Bill had left in various stages of completion. This group, consisting of Sarah Cole, Kate Devlin, John Drury, Artie Fleischer, Jennifer Rock, and Scott Swann, played a vital role in making a first attempt at a readable manuscript. Any teacher could feel only pride at leaving a legacy of such bright, dedicated, and energetic students.

The following winter, Kate Devlin undertook the enormous task of tracing the sources of Bill's many quotations and assembling an initial bibliography for the text. In addition, her cheerfulness and enthusiasm kept me going through what I began to realize was a truly daunting task. Ernst Mayr and Ian Nisbet both read drafts of the manuscript at this stage and made many useful suggestions for revision. Perhaps most important, Ernst Mayr suggested that we generally shorten the text and leave out the partially completed chapters on evolution. Ian Nisbet also read later drafts and, while largely disapproving of our efforts, provided an invaluable reality check to the process.

Upon Kate's departure for and Sarah's return from graduate school, Sarah Cole took over as chief editorial assistant and firmly but gently cut the revised but still overly wordy manuscript by half. In addition, Sarah pursued the remaining references, compiled species lists, prepared a table of contents, wrote countless letters, and scoured Bill's files for key pieces of the final chapters. Without her dedicated service the book would never have passed the first draft stage.

Once we had a clear document to work with, a second "Book Bunch," which consisted of Sarah Cole, Mary Drury, and several of Bill's colleagues, including Richard Borden, Craig Greene, Ed Kaelber, and Steve Katona, met regularly to discuss chapters and make further corrections.

I am grateful to each of these individuals for their patience and kindness. I particularly thank Mary Drury for providing the space, time, and trust to let the project happen. Cynthia Borden-Chisholm drafted figures, and Mary Drury searched through Bill's sketchbooks, drawings, and slides for additional appropriate illustrations. J. K. Anderson read several early chapters and made helpful suggestions as to the role of an editor. I am very grateful to Lou Rabineau for giving me a leave of absence to work on the manuscript. I also thank Karen and Clare, who put up with a distracted husband and father as this stepchild was developing.

At the final stages of editing Ernst Mayr once again stepped in with excellent advice and suggestions for publication. I am sure that Bill would have wanted special recognition of Dr. Mayr's importance as teacher, mentor, and friend, but I personally cannot thank him enough for his role as "fairy godfather" in bringing us to a point where we could go to press.

In summary, many people have come together to do part of what Bill had hoped to do. The book can never be what he—or any of us—wanted it to be, but I hope that at the least I have not traveled too closely in the steps of Kipling's Disciple:

"It is His Disciple
Who shall tell us how
Much the Master would have scrapped
Had he lived till now.
What he would have modified
Of what he said before.
It is His Disciple
Shall do this and more . . . "

Introduction

How, in an age of sophisticated mathematical models of complex inter-active systems, should I start a book which argues that nature works on the basis of one-on-one species interactions, variability, and chance? The job needs doing because I am uneasy with the repeated assertions that nature's norm is balance, that this balance is fragile, and that current human activities invite the collapse of entire, complex ecosystems. In contrast, what I have seen during decades of fieldwork is neither pervasive order nor chaos, but comfortable disorder. I offer an alternative model of how plants and animals interact and how humans think and behave like other animals. I will try to show that disorder is what makes the natural world work.

I have the nagging conviction that we humans can better deal with the difficulties that face us if we identify the problems accurately, learn what causes them, and apply creative solutions. I believe that clinging to romantic notions of nature's grand design only saps the strength of the conservation movement. I present an alternative notion because most of the environmental movement has ignored or rejected the changes in thinking that have infiltrated ecological theory since the mid-seventies.

THE ORIGINS OF THIS BOOK

My interest in the outdoors began early. As a small boy, I was strongly influenced by "birding" in the fields, woods, marshes, and brooks, and on the seashore of Rhode Island. Roger Peterson's *Field Guide to the Birds* allowed me to recognize all the birds I saw, in contrast to the much more painful experience that I had with identifying plants and insects.

While I was birding, I was eager to find new species but I was also preoccupied with drawing. Drawing required that I watch individuals closely in order to recognize them by their personality and put that down on paper: their proportions, how they moved, how they perched, how they scolded, how they expressed alarm, how they flew. From this I formed my own pictures of them and how they interacted with their world. In high school, as luck had it, I was excused from organized sports. All I had to do was walk a six- to eight-mile route birding along the beach, across the farmed point, the open marshes along the reservoirs, and the hundred-acre woods. I had been a loner, but in England, where I went for a year on an exchange scholarship, I learned that lots of people shared my interests.

In my last year as an undergraduate at Harvard I took a course in ecology. At the time, ecology was largely physiological in focus, involving descriptions of large-scale ecological systems and the climatic parameters identified as determining them. The whole thrust of this ecology was order, cause-and-effect relations, and equilibrium conditions. The course had little direct application to what I was interested in, and the clash of interests became even starker during my oral examination as a candidate for honors. In my thesis I had developed the idea that the particulars of species behavior are as important as morphological structures in providing species recognition. I was berated for the suggestion by a biochemist who called the topic and ideas old-fashioned. He threatened to fail me unless I could describe the structure of the active element in the blood protein hemoglobin. Fortunately, I could and did so. This vividly illustrated how differently people see what is important and what understanding consists of.

When I came back to graduate school after World War II, I wanted to continue my study of birds but was told, in no uncertain terms, that at Harvard the only acceptable study of bird behavior would involve laboratory studies of endocrinological mechanisms or the physics and anatomy of sound detection. So I shifted my focus in order to study geographic botany with Hugh Raup. Raup worked outdoors and in the north, and I could learn from him about the structure of vegetation, which, after all, provides birds their habitats. Raup made it immediately clear that understanding patterns of vegetation depended on the geology of unconsolidated materials. This led me to study the geology of Pleistocene deposits, and the combination led to a job that required me to run wild rivers in Alaska to describe the floodplain terrain in geological and botanical terms. I embraced the work with considerable enthusiasm. When a Calvinistic biochemist challenged me for doing this research because I enjoyed the outdoor work, I answered, "But of course. Why else?"

A job directing the education and scientific staff at the Massachusetts Audubon Society created the chance for me to set up a field station and work for conservation. For twenty years I turned my best efforts to finding and using currently available ideas about animal behavior and population biology, in the service of conservation of birds and habitat in southern New England, and to raising the quality of amateur field studies. This brought me into close cooperation with people who shared my appreciation of, enjoyment of, and gratitude toward nature, and it brought me into confrontation with constituencies holding very different perceptions of nature.

Some of the most helpful contributions to conservation biology came from animal behavior and the renaissance of Darwinian natural selection. The Darwinian centennial in 1959 set off a flurry of studies that reexamined the meaning of natural selection and clarified for me the fact that the process changes the rules of behavior at the level of the individual.

Plate 1. Bill Drury bird-watching with his eldest sons, Billy and Tim, in 1957. Photo from Drury family archives.

Interestingly enough, in spite of Darwin's emphasis on the importance of individual organisms, the ecological establishment seemed to be continually drawn to theories invoking higher levels of selection based on the theory that nature is made up of orderly entities called communities. An equally self-assured conservation movement based its rhetoric on "nature's balance." These people thought of nature as it "ought to be"—the product of years of development free of disturbance.

Equilibrium theory, the characteristic ecology taught in introductory textbooks, clearly provides the intellectual foundation of politically active environmentalists. I must acknowledge that these traditional principles have provided powerful weapons in the political struggles to serve the interests of wildlife, and that for me to presume to discredit them is tantamount to heresy. But these ideas can become doctrinaire, and many practitioners have been both sanctimonious and anti-intellectual. Such excesses have paralyzed attempts to pursue conservation as an applied science and have justified a powerful backlash by equally contentious and doctrinaire groups who argue from the same tenets that nature was created for human betterment.

The people involved in many of the controversies that I will discuss have scarcely been able to communicate with each other because their worldviews contrast so fundamentally. Moreover, these controversies are not resolved by reference to evidence and logic. Eventually the proponents and opponents die off and the members of the next generation cannot figure out what the fuss was all about. To some extent, these differences in attitude may be part of the inner natures of those involved.

DEEP PROGRAMS

I think that we can find evidence that these patterns have been to an important degree programmed into us. I don't think we have recognized the degree to which programming in our youth can affect our thinking in later years. This programming affects our scientific attitudes most pervasively. We surely did not recognize when we were being programmed—it proceeded in the most innocent circumstances, in the early years when we were beginning to learn a new discipline and were overwhelmed most of the time.

Remember the lessons that you were taught in a foreign language, chemistry, or geometry? The first terms were filled with memorizing a welter of disconnected, new information. The details floated in a kind of limbo, so you had to create jingles to remember them. Then, after several terms, the learning became much easier. I think that the change occurred because you had created a new scaffolding in your mind, a system into which you could fit the new things that you were called on to learn. I call that scaffolding a deep program. While you were creating the program, you entered ancillary material inherent in the scaffolding, and you were not conscious of it. You were vulnerable to what, if it occurred in other circumstances, might be called brainwashing.

You were not aware that while you were being taught chemistry, you were also being taught deterministic Newtonian reasoning, the assumption that atoms and molecules exist in stereotyped forms and will behave in exactly the same way regardless of sample; the changes that occur as concentrations increase will be exactly reversed as concentrations decrease. Later on it is difficult to know when you are thinking in ways that were programmed into you, and when you are thinking for yourself. The programs were put in places so deep in your mind that you cannot bring them up and consider whether you really want to think that way. I'm convinced that such programs seriously limit the way we think about nature and ourselves. The immediate and long-standing acceptance of the closed system models of W. M. Davis, Braun, and Clements are examples of this phenomenon. Thousands of naturalists and ecologists have become convinced of the validity of these ideas because they were not introduced to the alternatives and did not have a chance to formulate their own theories during a childhood outdoors.

At first glance many of the questions that I will raise throughout this book might seem to be limited in scope. The geology and flora of the Appalachians, extrinsic or intrinsic regulation of populations, and different feeding systems in birds may intrigue the specialist but have little practical application to our everyday lives. Viewed in terms of deep programs, however, the effect of deterministic ideas in ecology may be profound.

PRESENTING THE ARGUMENT

I am going to be talking about the kinds of order found in natural systems and the models contrived to help us understand them. I will be drawing your attention to both the "is" (what you can go out and see for yourself) and the "ought to be" (theorems of traditional ecology).

A first principle is that chance and change are the rule, the future is as unpredictable to other organisms as it is to us, and natural disturbance is too frequent for equilibrium models to be useful. "I returned, and saw under the sun, that the race is not to the swift, nor the battle to the strong, neither yet bread to the wise, nor yet riches to men of understanding, nor yet favor to men of skill; but time and chance happeneth to them all" (Ecclesiastes 9:11).

At the heart of my argument is the idea that while order and predictability exist in physical systems like ocean waves and rivers, individual organisms facing an uncertain future cannot afford consistency. Because the world is unpredictable in space and time, parents that produce diverse offspring will be better represented on average in the next generation than are those that produce uniform batches of young. Variation and changing environmental conditions select those individuals who produce what might appear to be a great excess of young. The fate of this excess provides the basis for the ecology of many other species.

The differences in viewpoints of ecology are so pervasive that I need to describe a wide sweep of landscapes and gossip about a lot of plants and animals to present my alternative. In the course of my argument, I will describe the deterioration of order from the mindless consistency of a meandering stream, through the marginally predictable behavior of plants and animals serving their self-interests under changing conditions, to the human potential for creativity, charity, and compassion.

My chapters are sketches that illustrate the atmosphere and interrelations between parts of ordinary scenes. In this they are like the *pochades*, or preliminary sketches in oil, painted by the Impressionists. The Impressionists used light and dark, cool and warm colors to represent a time of day and quality of the air, such as a hot summer evening after the rain. The Impressionists reached across barriers that usually separate arts

Plate 2. A common tern. Photo by the author.

from sciences. In both painting and science, as V. A. Howard said, symbols shape the form of our thought. Both, after all, depend on accurate observations of nature and creative interpretations of the patterns evident to those who will notice them.

In the following chapters I will describe the interplay between the complementary colors that controversies create, and the jarring effects when abstract theory (heavenly order) clashes with field data (earthly chaos). I will use sketches of landscapes and species as spots of color; they are not "the whole story." It is up to you to decide whether they help you understand and appreciate what you see outdoors.

I will focus on certain themes throughout the argument and examine how classical ecological theory uses these themes. These are mostly those loosely defined concepts that litter ecological theory, such as resource partitioning, trophic levels, competition, carrying capacity, equilibrium, succession, and so on. When used as very general terms, they provide pigeonholes for our thinking and imply self-organizing principles that work in their surroundings to increase order among the organisms to which they are applied. I think that we must examine each of these and test them against the idea of minimum complexity and against what evidence can be gathered from the field. As Theophrastus said, order should not be presumed in nature; it must be demonstrated.

SOURCES

Braun, L. 1947. Development of the Deciduous Forests of Eastern North America. *Ecological Monographs* 17 (2): 213–19.
———. 1950. *Deciduous Forests of Eastern North America.* Philadelphia: Blakiston.
Clements, F. E. 1916. *Plant Succession: An Analysis of the Development of Vegetation.* Carnegie Inst. Washington Publ. No. 242:1–512.
———. 1936. Nature and Structure of the Climax. *Journal of Ecology* 24:252–84.
Davis, W. M. 1899. The Geographical Cycle. *Geography Journal* 14:481–504.
Howard, V. A., and J. H. Barton. 1986. *Thinking on Paper.* New York: William Morrow.
Theophrastus. 1916. *Enquiry into Plants and Minor Works on Odours and Weather Signs.* Translated by A. Hort. Loeb Classical Library. New York: G. P. Putnam's Sons.

1 Natural Curiosity

Interest in natural history is old. For hundreds of centuries, lucky individuals have had their lives enriched by their interest in the world around them. In the beginning, when our ancestors were hunters and food-gatherers, this curiosity had practical value. Early humans had to accept nature as it was—to learn about kinds of animals and plants: which were which, where they grew, where they didn't, what was good to eat, what was dangerous; as well as where to find water, where to cross a river, where to find shelter, how to backtrack home, and how to return to a specific place another year.

One of the most powerful advantages of mobility and an expanding brain was surely the ability to search over a large area and to know it, to recognize important landmarks at a glance, and to tell your friends about them. It must have become important very early on to explain whether the rewards were worth the effort. This led to counting: how far, how many. So numbers are very old and at one time were more sophisticated than our own plodding, linear ideas of quantification. The methods of "primitive" people, as evident among older Inuit and Da-nene people of the north, emphasized mainly the significant figures and "natural logarithms." Ten or twelve is recognizable from six or eight, but sixty-five isn't worth separating from seventy-five. The larger the numbers, the coarser exponentially the grid.

Far back in our history humans showed a deep-seated need for predictability. People will accept uncritically the most bizarre forms of contrived order. During early times people repeatedly had to decide which way to turn, how to anticipate the next moves of their quarry, and where fat roots grew, and hence they wondered about deeper questions that we now call the biology of behavior. Some got ideas, had luck, and became skillful hunters and gatherers. Social status undoubtedly affected how rapidly ideas spread, and that status was measured as much by creative excuses as by food brought home. Those who could create artful explanations subject to several interpretations became shamans or priestesses. Now they are called economists and ecologists.

In this chapter I will present some of the philosophical background upon which much of what we now regard as ecology was built. I will begin with the Greek philosophers and work my way to the present, tracing how these early ideas became incorporated as often unstated assumptions of science. I will conclude by discussing the implications of this work on our understanding of human societies and other organisms.

PHILOSOPHICAL ORIGINS

I expect that during most of history people dealt with uncertainty the way we do now, by analyzing what they could control and mystifying what they could not. People sought order then as now by deceiving themselves and contriving order where they could see none, but there are some important differences. What little order so-called primitive people know, they attribute to demonic beings or spirits who represent and govern nature. This is very different from seeking pervasive order, which emerged when economic and cultural institutions removed influential thinkers from direct, daily contact with nature.

Granted the size, complexity, and confusion of this world, thinking about the future scared people. So the ancients categorized their world by parts. Homer, in his *Iliad*, showed the importance of belief in direct communication with supernatural beings. Agamemnon explains why he stole the slave girl from Achilles: "Not I was the cause of this act, but Zeus, and my portion, and the Erinyes who walk in darkness: they it was in the assembly put wild ate upon me on that day when I arbitrarily took Achilles' prize from him, so what could I do? Gods always have their way" (*Iliad*, bk. 19). Achilles accepted the explanation, for he also was obedient to his gods (Jaynes 1976). People of Homer's Aegean Sea believed in the reality of voices that they heard. This anecdote illustrates the impressive impact of belief systems on a person's perception of reality.

The magical poetry of priests and prophetesses dominated people's minds during many millennia. A major change in thought occurred in the fourth and third centuries B.C. among urban thinkers of classical Greece, and the results of this change dominated Western intellectuals for about fifteen hundred years. Greek philosophers developed a "religion" of analytical thought, and a central topic of their analysis was the nature and origin of order, as manifested on earth and in the universe.

The insulation from the confusion of undeveloped areas provided by cities allowed Greek thinkers to focus on abstract principles and absolutes. Socrates, a confirmed townsman, commented that "fields and trees will not teach me anything, but men do." This preoccupation with absolutes is noted by Stone (1989): "For Socrates, if you couldn't define something with unvarying comprehensiveness, then you didn't really know what it was."

Another philosophy was embraced by those who believed that the constancy of atoms as ultimate components imposes itself upon all else. We would call these thinkers reductionists.

An alternate view of nature and order was that certain forms provide the only realities, while our perceptions are less-than-perfect manifestations of the forms. Plato, immediately after Socrates, is usually credited with originating this belief. The naturalist Aristotle contributed ideas of considerable influence during the succeeding centuries. Among Aristotle's thoughts was the idea that all living structures and activities have biological meaning, or as we would say now, survival value—so the roots of Darwinism are old.

The concept that the earth is an imperfect manifestation of a higher order has the fringe benefit that the cognoscenti can dismiss those who don't recognize the order, by claiming they lack sophistication. In separating themselves from the clutter of the world, Greek intellectuals pioneered the use of mental constructs to simplify and contain the big, complex, variable, confusing, and unpredictable world we live in. They developed the logic of reasoning deductively from assumptions, and they defined the rules of modern philosophy and science. In so doing, though, they started the trend to narrow the attributes that academics recognize as intelligence. Emphasis on ability to internalize definitions of general principles and facility at deducing the behavior of corollaries has eliminated several categories of intellectual talents that have been valuable in the lives of those whose preoccupation is not primarily with abstract concepts. I am confident that, if released fifty miles out to sea, most of my commercial fishermen friends would quickly find their way home, but I doubt that I would ever hear from some of my academic friends again.

Socrates and Plato concentrated on constructs of the human mind. They believed that dialogue, the interactions between different ideas, would reveal understanding. "Truth" could be discovered through human discourse. Aristotle, a biologist, made what are for me some major contributions that later philosophers have failed to consider or maybe to understand. Aristotle proposed that there are in nature some realities for which there are not equally viable alternative understandings. If one starts with false premises, dialogue and logic will not necessarily lead one to correct conclusions.

During the Middle Ages, kings benefited by deception about their "divine right" to the throne. This myth projected fear into the hearts of those who might have wanted to seize power, and it gave the kings' people the security of knowing that supernatural forces were on their side. Acceptance of authority was central to the discipline of the medieval church where scholars gathered. We should not be surprised, then, that medieval scholars answered the question of how many teeth a horse has by reading Aristotle, not by going to a stable and looking for themselves.

Newton, basing his ideas on isolated, internally consistent systems, showed that nature was governed by simple, basic forces: mass and movement. The resultant scientific reductionism asserts that the more consistent and orderly sciences provide the ordering principles for understanding all science. Many reductionists believe that only when systems can be reduced to quantitatively formulated order are they understood. As long as everyone accepted these as explanations, they worked.

According to the mechanistic philosophy created in the seventeenth century by Galileo, Descartes, Bacon, and Newton, the parts of Nature's Plan functioned like a great, integrated, well-maintained machine, and thus essential processes can be defined rigorously in Cartesian mathematical terms. The philosophy allowed scientists to think of individual organisms as mindless cogs playing a part in a large organized entity. This mechanistic view depends on assumptions of consistency in the structure and operations of nature (the fixity of Creation). Similar convictions have led many to believe that the only alternative to real order is real disorder or chaos.

We can see that the philosophy of Galileo, Descartes, Bacon, and Newton has had a massive impact on the philosophy of science. The orderliness of physical systems, and especially the assuredness and rigor of classical physics, has haunted field biologists for the last two centuries.

THE INFLUENCE OF LINNAEUS

Linnaeus's essay "The Economy of Nature," written in 1749, offered a model of nature that defined many attitudes still generally accepted. As Frank Egerton said: "Linnaeus defined the concept (the Balance of Nature) and tried to make it the foundation of an ecological science" (1973). In the Linnaean system, every species has its allotted place and assigned work in the general economy.

Linnaeus had a powerful influence on field biologists because his major contribution, *Systema Naturae*, provided a very usable, quantitative sys-

Purple flowering raspberry, pen-and-ink botanical drawing, 1941.

tem for classifying living things. His system spread widely because it worked and could be readily understood, and, furthermore, because it provided a standard, simplified system of scientific names: a genus and a species.

Linnaeus's system carries a profound subliminal message: that each species was created as it is. This Linnaean "type concept" had a strong and pervasive influence on the thinking of field biologists. It was a message that most naturalists at the end of the eighteenth century wanted to hear, and the message came at a critical time, the start of the great age of exploration. The late eighteenth and the nineteenth century were periods of meteoric growth of the art of classifying the collections brought back to the museums of Europe. The idea that species are fixed justified their classification according to consistencies in morphology within populations.

During the first part of this age of classification, most written material came from the pens of closeted individuals working on very few specimens taken at great distances from each other. Few curators had opportunities to travel even in their youth. Under these conditions it was easy to confirm one's preconceptions about discrete and immutable species. Then, at the beginning of the twentieth century, the type concept was applied to the classification of plant communities. The concept of the Linnaean, morphologically defined species persisted in botany into the 1950s. Classification focused on the "typical form," and population genetics focused on "population norms." I remember my botany professor searching over a hillside covered with individual plants of a particular species until he found a "perfect" representative of the variety he had in mind. His comment when he finally found the right specimen was that this individual was "typical."

THE ROMANTICS AND THE ORIGINS OF HOLISM

During the early decades of the nineteenth century, the limited alternatives of Linnaean preoccupation with dead specimens and the Bacon-Newtonian preoccupation with nature as a well-organized machine convinced most scientists that species' lives consisted of appointed rounds as the wheels turn in nature's machinery. This had important philosophical implications. Many field naturalists were dissatisfied with the confining atmosphere that left out "aliveness" and "appreciation."

Alexander von Humboldt, considered to be the founder of plant geography, admitted to having as much appreciation for the beauties of sublime landscapes as interest in his scientific studies, and he established himself as a pioneer romantic natural philosopher. By midcentury a group of dissatisfied field naturalists carried the theme of appreciation and emotion to extremes.

Johann Wolfgang von Goethe and Henry David Thoreau, alienated by the scientific natural philosophy available at the time and responsive to the aliveness of living things, sought explanations for the soul of living nature. The idea of aliveness, *anima*, became associated with pagan *ani-*

mism, that is, the soul in an oak tree or the spirit of a bear. This was Romanticism, a form of denial of the ugliness of the times.

Just as the Industrial Revolution remains a major manifestation of the power of mechanistic science, so its pollution, drabness, and exploitation disillusioned many poets. The poets rationalized that God had made nature, but humankind, in making the cities so ugly, had lost sight of Him and had damaged Nature by disturbing Her balance. Hope for humankind lay in restoring oneness with nature, because by this reconnection people might regain touch with all-pervasive spirituality. By the end of the nineteenth century the French philosopher Henri Bergson had identified an "élan vital," or "life thrust" (Worster 1977). This driving force of life would ensure progress in evolution.

It is too facile to dismiss vitalism as a mystical alternative to materialistic science. The nineteenth century was a time of belief in universals; the machinery of nature was already a common metaphor. It was no more irrational to conceive of an all-pervasive "life force" tying together living systems than it was to conceive of an all-pervasive "gravity force" tying together material bodies or an all-pervasive "ether," through which light waves were then believed to be propagated.

Darwin's natural selection provided the means of change, within which the mechanism of competition provided the wedges or leverage for progress. Using a metaphor of the individual organism's development to maturity , Ernst Haeckel coined the aphorism "Ontogeny recapitulates phylogeny," which suggests that in its personal development each individual passes through the stages of evolution.

ECONOMY OF NATURE

Thus toward the end of the nineteenth century several materialistic mechanisms were suggested to explain progress and evolution, which took the place of the Divine Plan and Creation in supplying purpose. From economics, biology incorporated Adam Smith's metaphor suggesting that in a capitalistic society an "unseen hand" guided entrepreneurial endeavors to coordinate the individual's self-interest. Those endeavors operating at a great distance and suffering from extensive delays in delivering returns would be replaced by ones near at hand and that brought returns more quickly. Social scientists have changed this rather straightforward statement to conform to the ideal of improving the operation of society. As Visvader and Borden put it: "The apparently chaotic actions of individuals, each concerned with their own advantage, would be guided as if by 'an invisible hand' (in the words of Adam Smith) to produce a harmony greater than any that could be produced by action deliberately construed to produce harmony" (1985).

Once ecology became associated with the "economy of nature," the unseen hand would play an important role in the assumed regulation of equilibrium communities. Late-nineteenth-century Victorian society also changed Darwin's concept of selection to be more consistent with its own beliefs, that selection and adaptation worked for the benefit of the species. Thus populations would develop adaptations to produce large numbers of offspring to compensate for population crashes, or they would have mechanisms to regulate population size to avoid stressing resources and inviting crashes.

MODERN TRADITIONAL ECOLOGY

Early in their history ecologists divided themselves into synecologists, or students of systems, and autecologists, or students of individual species. When I was introduced to these two disciplines both were basically physiological in approach. Students of systems found it convenient to consider their subjects self-sufficient entities. This led to holistic concepts of plant and animal communities and justified study of the forces that move events in the "system": energy flow, nutrient cycling, community structures, and the processes that confer stability. Autecologists focused on physiological ecology, the adaptations that allow individuals to meet the challenges of the physical world. Neither looked at how the organism actually lived its life.

Let us look specifically at how some historical and philosophical concepts have been incorporated into modern ecology. I shall discuss most of these ideas again in future chapters, but it may be helpful to look at the parallels while the history is still fresh in your mind.

Three ancient ideas have been important to the development of modern ecology. Aristotle's concept of Microcosm/Macrocosm is that the internal organization of the parts mimics that of the whole, hence subsets of a large system should stand by themselves. His concept of the Great Chain of Being includes the traditional separation of "lower" from "higher" forms of life and the conviction that the differences are progressive. His concept of the Balance of Nature is evident in the nearly universal use of equilibrium models for natural communities.

Microcosm/Macrocosm

The concept of natural communities as cooperating wholes may simply be a hopeful anthropomorphism based on the socioeconomic structure of Greek city-states. In contemporary use, "community" stands for a set of processes or rules that govern groups of interacting species. A central but unspecified tenet is that natural processes (from God's Creation, through an "unseen hand," to natural selection) bring out harmonies among coevolved community members.

The concept of a natural community as a microcosm has been called "the first important theoretical construct to develop in American limnology" (Hutchinson 1963). The idea was first specified by Forbes: "The lake is an old and relatively primitive system, isolated from its surroundings. Within it matter circulates, and controls operate to produce an equilibrium comparable with that in a similar area of land. In this microcosm nothing can be fully understood until its relationship to the whole can be clearly seen" (Forbes 1887 in McIntosh 1985).

Note the emphasis on both the *isolation* of the lake and the *equilibrium* that will be maintained. From an understanding of the microcosm (community) of the lake it is assumed we can derive understanding of the macrocosm of the wider world. Because the one appears to be at equilibrium, we assume a similar case in our studies of the other. We thus find two of three Aristotelian principles—Microcosm/Macrocosm and the Balance of Nature—already assumed as part of ecological writ.

Allee, Emerson, Park, Park, and Schmidt made the functioning community explicit in their book that became the standard American ecology text during the 1950s and strongly influenced texts that have appeared since then: "The community maintains a certain balance, establishes a biotic border, and has a certain unity paralleling the dynamic equilibrium and organization of other living systems. Natural selection operates upon the whole inter-species system, resulting in a slow evolution of adaptive integration and balance" (Allee et al. 1949). Again, we see an emphasis on both separation into parts and an internally maintained balance or equilibrium.

The Great Chain of Being, Progress, and Succession

Much ecological theory includes the assumption that natural ecological change is progressive, that the end point is an equilibrium community that is the fullest development of local ecological conditions. The process has been called succession. The organization and gradation found in successional theory harkens back to an earlier idea: that of the Great Chain of Being.

Plenitude, continuity, and gradation are the primary principles underlying the Great Chain of Being (Lovejoy 1978). For Plotinus (A.D. 205–270), the first principle, plenitude, meant that all possible things ought to be. Hence, if all things were equal, all things could not be. Therefore the gradual steps in the Great Chain of Being are necessary. Saint Thomas Aquinas (1125?–1274?) developed this principle further into one in which the earth must be full of all possible species and must be the best of all possible worlds. To suggest otherwise would be to doubt the goodness and wisdom of the Maker.

Successional theory is based on an additional presumption that differences in communities observable in space at one time are equivalent to changes that take place over time; for example, zones of vegetation around a pond or among sand dunes are stages in the development of vegetation toward a stable community. It seems to involve a more pervasive, hidden assumption: that the maturation of an organism through stages of development can be projected onto the interrelations between organisms. Odum, one of the chief proponents of the concept of succession put it this way: "As viewed here, ecological succession involves the development of ecosystems; it has many parallels in the developmental biology of organisms, and also in the development of human society. The ecosystem, or ecological system, is considered to be a unit of biological organization made up of all of the organisms in a given area (that is, "community") interacting with the physical environment so that a flow of energy leads to characteristic trophic structure and material cycles within the system" (1969).

The Balance of Nature—Equilibrium

Even as the concept of communities may have been derived from the ideal of the Greek city-state, the idea that mature ecological systems maintain equilibrium—a condition of "harmonious interactions"—may be a hopeful abstraction from ancient religious beliefs in the eternal order and ultimate fixedness of God's creation: the natural state is "good" and the product of design. Bormann put this in technical language as the achievement of balance between input and output: "It seems to me there is an indisputable amount of evidence to indicate that on a particular site, following some type of disturbance, there occurs a sequence of vegetational types and a developmental sequence that leads, we think, toward stability. The type of stability which is being approached is one wherein community respiration is approximately equated to community photosynthesis. The two processes tend toward balance, and thus we have a truly stable system" (1969). Odum stated explicitly that natural systems close themselves as they mature: "An important trend in successional development is the closing or 'tightening' of biogeochemical cycling of major nutrients, such as nitrogen, phosphorus, and calcium . . . " (1969).

Note that equilibrium theory discounts physical forces such as fire, rainstorms, high winds, and geological processes. Successionists consider these "disturbances." But, if plant roots could indeed restrain all soil erosion, rivers would carry no sediments and soil creep could not erode the valley sides. An intense downpour may occupy all of a small drainage basin and set off flash floods and mud slides. Such rare events of exceptional impact move unconsolidated soil materials downslope and downriver.

THE ALTERNATIVE VIEW: OPENNESS IN ECOLOGY

An alternative metaphor for a natural ecological system is provided by a river. The form of a river is created and maintained by the movement of energy and materials, water and sediments, through it. It is an "open system" in the sense of von Bertalanffy (1950), Chorley (1962), and Drury and Nisbet (1971). This metaphor also suggests a reanalysis of teleology, or goal orientation, as applied to the system. Is a river "flowing to the sea," or is it responding simply to the local gradient in its bed and sediments delivered to it from the side slopes? In the interactions within communities, do internal processes such as competition move species aggregations toward an equilibrium condition or does each species respond independently to the quality of the site and the activities of its immediate neighbors?

Inherent in this is the question of whether all species present are essential to a community's operation as a holistic entity, or whether the units of action are nexuses of species that have strong effects on each other while most species are only weakly involved or irrelevant. Robert Paine made the issue clearer in identifying "keystone species," such as the predatory starfish that eat almost everything that lives below high tide (1969). He showed that the removal of starfish results in major changes in local species composition.

Openness of ecological communities, unevenness of habitat quality, and ability to disperse are truly important bases of species persistence. Heterogeneity and redundancy of habitat patches provide the infrastructure of natural ecological systems, rather than stability and homeostasis.

My rejection of traditional ecological beliefs is a thesis running throughout this book. A strong tendency to accept the existence of self-organizing principles as inherent in natural systems has been evident in the development of ecological theory since the start—that is, balance of nature, community, the adjustment of birth rates to death rates, succession, density dependence, and so on.

In contrast, I feel that ecosystems are largely extemporaneous and that most species (in what we often call a community) are superfluous to the operation of those sets of species between which we can clearly identify important interactions. Individuals and species do not benefit as much by becoming integrated into a functioning system as they do by maintaining options of disengagement so that they can crop excesses in the good times and move out during a crunch.

Why do we continue to imagine that self-organizing principles and holistic benefits exist in natural "communities?" I think that the problem is buried deeply in the human psyche. Most of us would rather put our confidence in supernatural (self-organizing) processes than address awkward realities that require us to acknowledge the consequences of self-serving activities.

SOURCES

Allee, W. C., A. E. Emerson, O. Park, T. Park, and K. P. Schmidt. 1949. *Principles of Animal Ecology.* Philadelphia: W. B. Saunders.

Bormann, F. H. 1969. A Holistic Approach to Nutrient Cycling in Plant Communities. In *Essays in Plant Geography and Ecology,* edited by K. N. H. Greenidge, 149–65. Halifax: Nova Scotia Museum.

Chorley, R. J. 1962. *Geomorphology and General Systems Theory.* U.S. Geol. Survey Prof. Paper 500-B. Washington, D.C.: USGPO.

Darwin, C. 1859. *On the Origin of Species by Means of Natural Selection.* London: Watts.

Drury, W. H., and I. C. T. Nisbet. 1971. Inter-relations between Developmental Models in Geomorphology, Plant Ecology, and Animal Ecology. *General Systems* 16:57–68.

Egerton, F. N. 1973. Changing Concepts of the Balance of Nature. *Quarterly Review of Biology* 48:324–50.

Haeckel, E. 1906 *The History of Creation.* 4th ed. London: Kegan Paul, Trench, Trubner.

Hutchinson, G. E. 1963. The Prospect before Us. In *Limnology in North America,* edited by D. G. Frey, 683–90. Madison: University of Wisconsin Press.

Jaynes, J. 1976. *The Origin of Consciousness in the Breakdown of the Bicameral Mind.* Boston: Houghton Mifflin.

Linne, C. A. 1751. Specimen Acaclemicum de Oeconomia Naturae. *Amoenitates Academicae* 2:1–58.

———. 1767. *Systema Naturae.* Stockholm: Impensis Direct.

Lovejoy, A. O. 1978. *The Great Chain of Being.* Cambridge: Harvard University Press.

Mayr, E. 1982. *The Growth of Biological Thought.* Cambridge: Harvard University Press.

McIntosh, R. P. 1985. *The Background of Ecology.* New York: Cambridge University Press.

Odum, E. P. 1969. The Strategy of Ecosystem Development. *Science* 164:262–70.

Paine, R. 1969. A Note on Trophic Complexity and Community Stability. *American Naturalist* 103:91–93.

Smith, A. 1937. *An Inquiry into the Nature and Causes of the Wealth of Nations.* New York: Random House.

Stone, I. F. 1989. *The Trial of Socrates.* New York: Doubleday.

Visvader, J., and R. J. Borden. 1985. Education for a Livable Future. Paper delivered at North American Association for Environmental Education meeting, Washington, D.C.

von Bertalanffy, L. 1950. The Theory of Open Systems in Physics and Biology. *Science* 111:23–29.

Worster, D. 1977. *Nature's Economy.* New York: Cambridge University Press.

2 The Seashore

Throughout the book, I will make the point that explanations should be as direct and simple-minded as possible. In this chapter I will emphasize the importance of local physical conditions and complex, one-on-one species interactions. The combination may confuse some readers and lead others to wonder how they are going to be able to "organize" the nature they see around them so that they can comprehend it.

The first *pochade* that I will describe is the seashore, because that is where the interplay of physical and biological effects in ecology is especially visible. My major point is that physical forces have important influences on the patterns of species that grow in the intertidal and subtidal zones. The effects of physical forces are obvious where winter storm waves pound on rocks and rearrange sandy beaches. The central role of physical conditions is also evident on a much larger scale. Two conspicuous physical features of the coastal landscapes of New England are, in the north, the rocky shores with many islands washed by impressive tides and, in the south, the sandy beaches washed by much less imposing tides.

THE INTERTIDAL LANDSCAPE OF ROCKY SHORES

In the rocky intertidal of northern New England and Atlantic Canada, a falling tide exposes a usually predictable sequence of zones. At the top is a dark green stain, blue-green algae. The next zone is an off-white carpet of barnacles. Below the barnacles is a broad band of seaweeds, bladder wrack, and knotted rockweed. The seaweeds cover nearly everything down to a sharply cut lower border at spring tide. A flush of species, most conspicuously tufted Irish moss and the long-bladed kelps, appears below the low spring tide level.

The blocks of stone jarred off bedrock collect in jumbled piles, offering surfaces with many different exposures and containing catacombs of passages. Areas higher on the shore are exposed to the cold winds and the sun and are lashed by waves more than those lower down. Conditions change as tides ebb and flood. Some surfaces are pounded, some are washed by surges, some feel only the rise and fall of the sea. These factors combine to create a lot of habitat variety. Conditions may differ more over a distance of a foot than they do over a distance of several fathoms, and consequently, species sort themselves out into markedly different combinations over short distances while retaining a regional similarity.

Winter storms may tear patches of seaweeds and jar mussels, snails, and crabs off their holdings. Space to put down is at a premium, and getting established is critical. Animals and plants live out their lives in a context of shoving, elbowing, and accidents that leave them to drift at sea.

While the sea constantly threatens disaster, its currents provide a flux of immigrant colonists from all up and down the coast. The adult plants and animals, while largely fixed themselves, release large quantities of eggs and larvae into the sea to be carried far and wide by currents. The members of the intertidal landscape are part of an open system, and the patches from which residents have been torn are soon resettled.

TRILL
ZZZZZZZ

oik

eng

HOVERING TRILLING

"SHARP TAILED GROUSE" DANCE

ALARM   ALARM AT NEST

White-rumped sandpiper behaviors, pen and ink, 1961.

SONG WHILE PURSUING FEMALE

eng oik

PARACHUTING AFTER
DISPLAY FLIGHT

PLUNGING TO THE
GROUND AFTER DISPLAY

ATTACK

SONG TO INTRUDER

WING UP DISPLAY

DISTRACTION

Effects of Wave-Action and Species-Species Interactions

We describe what we see on intertidal rocks, and to simplify the task of communicating with others we identify many combinations of species as communities. At that point some unconscious assumptions intrude. The landscape appears to be constant because over the short period of our own experience we are not aware of changes in the seaweed-covered rocks. Our impression is one of consistency and continuity, as long as we don't look too closely. The unconscious assumption that the landscape is constant during the lives of the plants and animals has led to the general assumption that communities persist and that community structures and processes ensure the persistence. Ecologists have deduced that interspecific interactions such as competition supply the assumed order and that populations are "normally" stable.

Most of the interactions between species are simple and direct. The picture becomes complex chiefly because there are many different interactions. Many people expect simplified descriptions of nature and simple functional explanations for the patterns, like the claim that the zones of algae are determined by the penetration of light. At another extreme some people accept nature's complexity as unfathomable and enjoy the mystery. I think that the real world lies between the two and is more interesting than either.

Barnacles feed by combing particles of this abundant life out of the water, and they close their shells when exposed to air to protect themselves from drying out. To survive, they must be washed over by waves a couple of times a day; thus, the reach of the waves sets the general upper limit of barnacles.

The line marking the lower limit of barnacles reflects the action of several processes. First, attached barnacles provide good "holding ground" for the holdfasts of knotted rockweed and bladder wrack. Once rockweeds are established, their whiplash discourages the establishment of barnacles. Second, the presence of barnacles discourages the grazing of periwinkles, herbivorous snails that graze on small "seedling" algae. Where there are few barnacles, the snails create open patches where rockweeds can attach. The clumps of these seaweeds provide shelter for dog whelks, which prey on barnacles. On headlands where wave shock eliminates mussel predators, mussels are able to suffocate barnacles in the lower intertidal.

Periwinkle grazing is a major factor in seaweed distribution. The green alga *Enteromorpha* can become established only in the absence of periwinkles. By feeding on green algae, periwinkles release the red alga Irish moss. So where snails occur at high density in the lowest part of the intertidal zone, Irish moss dominates.

Periwinkles are limited at high-tide levels by drying. They must be able to reach damp shelter from the sun at low tide on hot days. In the midtide zone, wave shock and the periwinkles' distaste for brown algae limits their activity, and rockweed and wrack can grow in dense stands. At the lowest intertidal, they graze the rocks clean of almost all ephemeral green algae, leaving Irish moss and scattered plants of laver.

Dog whelks are sensitive both to drying and to wave action, so where dog whelks are knocked loose by wave shock or baked by the sun, barnacles persist. The whelks must retreat into the shelter of the fronds of weeds at low tide. The distribution of sheltering seaweeds thus establishes the upper limits of whelk distribution.

At lower intertidal levels, the dog whelk's preference for barnacles and mussels eliminates most of these species, opening space for wrack and rockweed. In wave-pounded midtidal areas, where wave shock jars dog whelks loose, mussels dominate.

Although blue mussels smother other plants and animals if given free rein, they usually occupy refuges that allow them sanctuary from whelks and starfish. Because whelks and starfish avoid pounding surf, they attack mussels only in sheltered places. In the lowest intertidal zone Irish moss dominates because both starfish and dog whelk prey heavily on mussels. If the predators are removed, mussels may become established and smother both Irish moss and barnacles.

Implications of Patterns in the Intertidal Zone

When I look closely at the rocks, I see no particular pattern in preference to others. The interactions are local, one-on-one, and change with time.

Paine said: "One of the more recognizable and workable units within the community nexus are subwebs, groups of organisms capped by a terminal carnivore and trophically interrelated in such a way that at higher levels there is little transfer of energy to co-occurring sub-webs" (1966).

The important species in a subweb include both those that would establish monopolies and those that frustrate the trend. I see the reactions of intertidal life to external forces in terms of patches of seaweeds and mussels that can be torn up in winter storms and the thousands of snails that will be displaced. I also see that the resilience supplied by the adaptations of the organisms—gross overproduction of young, mobility of intermediate stages, and ability to colonize bare places—soon makes the effects barely detectable. Hundreds of thousands of larval organisms produced in the spring settle and replace the ones that were lost. Predatory snails move in from the side, mussel spat settles. Crabs and starfish, the top predators, appear as if out of nowhere and the rocks seem to be the same again.

I argue that there are vertical interactions—between geological processes, plants, and animals—that help us to understand what is going on. There are interactions between physical forces in the sea, the surfaces and exposures of the rocks, and the distribution of the plants. In their turn the patches of plants reflect interactions with animals. The rocks do not simply supply substrate that plants occupy, and plants do not simply supply habitats for animals. Once seen, most of the interactions are simple and direct. Complexity seems to be a figment of our imaginations driven by taking the "holistic" view.

Many of the patterns evident in what is growing on rocky shores reflect wave action. The waves rush over rocks, pulling at the seaweeds and moving smaller stones so that everything has to hold tight or constantly swim to stay put.

Wave forms have consistent features subject to quantitative descriptions of forces and consequences of the tidy sort that delight scientists. As force is applied the wave increases according to a sinusoidal curve. The wave increases slowly at the start, then more rapidly, then more slowly as a graph of wave height approaches but does not reach an upper limit, the asymptote. This sigmoid curve and its mathematical description, the logistic equation, are relevant to essentially all situations of wave growth and equilibrium. A family of S-shaped curves can be prepared for the growth to equilibrium of water waves of different wavelengths and periods according to force of the wind, length of time the wind blows, and length of the fetch. Waves actually conform to these lines drawn on a graph.

Over a period of several days the wind blows from several directions and creates wave trains converging from different angles, each moving through the water independently. Little waves climbing over the flanks of big ones and big ones running in several directions create confused seas. Then the simple elegance of physical systems seems to deteriorate. Passage of a boat through the waves can be easy or tortuous depending on the way the waves from tide and wind move through each other, the artistry of the boat design, and the art of the person at the helm. What to one helmsman is confusion and discomfort, another transforms into a graceful dance, lifting and turning.

The regularities found in physical systems like those evident in the growth and achievement of equilibrium conditions in waves exhibit an admirable orderliness. Small waves behave just the same as large waves, as if transcending major changes in scale. While this is true in part for physical systems, it is misleading to use the behavior of small plants or animals or the interactions between a few species to predict the behavior of many species together.

HABITAT AND PLANT ASSOCIATIONS ON SANDY BEACHES

A similar relationship between physical order and biology can be found in plants of the sandy shore. The wind blows almost continuously on the seashore and moves sand across the beach. This is one of the most important forces active in the constant change of the shapes of beaches. Yet this continual change in detail must acknowledge an agelessness about the sandy beaches that embrace the seashores of the world. Changing on a short time scale and changeless on a long time scale is the mold in which has been cast the behavior of seashore animals and plants.

Most people are familiar with beaches only in the summer during the good times, when things are "normal." Under sunny skies and gentle breezes low waves collapse along the beach and move sand in the swash as they roll in. Their energy is dissipated before the waves wash back down again, and they leave the sand behind. So sand is gradually moved up onto the beach by fair-weather waves. But summer months make up barely a quarter of the year. People who have only a summertime familiarity with beaches can miss most of the action, as people who know migrating birds only on their breeding grounds miss most of what affects the lives of these species in other lands. An understanding of some of what goes on during the rest of the year is important, because the things that happen when most of us are not at the seaside indirectly create the patterns we see in the plants.

The winter beach is broad, wind-packed, often damp, a surface from which most of the fine sand has been blown. On top of the winter berm the beach forms a gently rolling surface, sloping upward over several tens of yards to the foot of the foredunes. On this upper beach the debris of winter collects: shells, seaweeds, the trash and dead skeletons of summer's strand plants.

Anyone who has walked a winter beach surely has seen sand blowing in sheets over the wet surface and collecting in windrows behind driftwood or in piles of seaweed. Coarse sand develops into a linear dune in the lee of a clump of marram grass. A fan of sand collects where wind slows as it is suddenly released after being confined between pieces of driftwood. The movement of sand is not limited to the outer beach, as you know if you have tented in the dune hollows. This continuous movement maintains the form and structure of sand dunes and dune hollows.

In summer relatively gentle breezes build ripples on the beach where the sand is blowing. Ripples are especially conspicuous on broad mounds of granite sand, where the darker, finer minerals (hematite, apatite, garnet) segregate into the hollows and coarser, lighter minerals (quartz) collect on the crests. The segregation results from the wind making sand grains jump and bounce along; the process is called saltation. When a grain of sand lands after a wind-born leap, it jostles other grains, setting them off to leap.

Vegetation of Sand Spits

Are the differences in vegetation you find walking the length of a sand spit beach the same as those you find walking a line through the dunes from the sea beach to the salt marsh? Sand spits build out over many years. According to the theory of succession you might expect the vegetation at the base, being older, to be more fully developed than that at the tip. Does the vegetation that you observe as you move down the spit appear to be earlier and earlier stages of succession, from places where plant processes have been active over many decades to places where plants have only recently been active?

It soon becomes apparent that the width of the beach and distance to the ocean make a big difference. In contrast, the length of the spit, whether two or ten kilometers, is not particularly important so long as the width remains constant. The vegetation in the foredunes and the outer dune slacks is the same all the way down the length of the spit. The same sequences—outer dunes covered with marram grass, slacks grown to beach heather, hollows covered with bayberry and cranberry, or aspens and pitch pine—occur at the base and at the tip. Sedgy and boggy hollows are bordered with bayberry and *Rosa rugosa,* both at the base and near the tip. Near the tip, where dunes are low and overwash is frequent, the contrast between the inner and outer dunes is lost, topographically and botanically.

The plants seem to respond to wind, blowing sand, salt spray, and depth to soil moisture—the current and local conditions. The taller, more diverse vegetation, such as thickets and forests, occurs at some distance from the surf zone and is not inversely correlated with distance from the base of the spit, as would occur if age played a part. At one extreme, exposure to flying spray, shifting sand, blasting winds, and periodic desperate drought is the lot of plants on the outer beach and outer dunes. At the other extreme, perennially damp hollows beyond the reach of wind and salt spray provide a haven with milder conditions.

It appears that plants growing in the outer dunes are more likely to experience movement of sand in a given fifty-year period than are plants growing in the inner parts of the dune system. The probabilities of wave overwash and dune blowouts determine how many years plants will

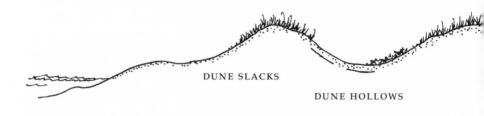

DUNE SLACKS

DUNE HOLLOWS

OCEAN   FORESHORE   OUTER OR FOREDUNES

Figure 1. Profile of a barrier beach. Courtesy of Cynthia Borden-Chisholm.

probably have in order to mature and fruit before disaster strikes. These probabilities control whether fast-growing wildflowers or slow-growing trees will be able to hold on. In this way, generation times are tested against the interval between invasions of waves of salt water or sand. Longer-lived plants are less likely to bear fruit before they are killed than are shorter-lived species. George Woodwell noticed similar effects around a source of radiation at Brookhaven National Laboratory in Massachusetts, and he assigned the differences to successional processes (1970). I am interpreting the differences more simple-mindedly, relating the differences to the amount of time available for plants to live out their lives.

Ranwell's study of winds in coastal dunes in Britain showed markedly less violent winds over the back dunes than over the foredunes (or outer dunes) (1972). Strong winds off the sea move coarse sand, piling it up among grass stems, but even in these places rain maintains a water table, though at considerable depth. Blowing sand may be excavated from around roots on the windward side of a dune, or it may bury and suffocate slow-growing plants on the inner slope. The roots of plants that live on active dunes must grow rapidly and opportunely. They must be able to withstand the strong winds that damage or kill tender tips. Storm winds carry salt spray inland and can have disastrous effects on plants already

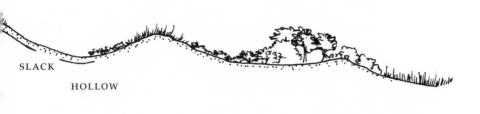

SLACK

HOLLOW

BACK DUNES   SALT MARSH

short of water and not adapted to salt stress. The distance to which storm winds disperse spray has an important effect on where species grow.

Describing Vegetation

Botanists describing vegetation have much the same purpose as the Impressionist painters. In order to communicate with others, they need to record their observations accurately and apply creative imagination to explain them. The student of landscape wants to describe the ordinary in the landscape. Observers, whether scientists or painters, cannot escape noticing some features at the expense of others. While the artist encourages this, the scientist usually tries to suppress it, but still the personal reaction is there.

Frank Egler said that vegetation is not equivalent to the list of plant species (that is, the flora) of an area but is made up of aggregations of species of plants (communities) whose visual and biological impact is largely the result of the growth form and relative abundance and dominance of the largest and most conspicuous plants (1977).

Botanists have found it easy to conclude that conspicuous species play important roles in the "community" they describe. Thus a bank of bushes in fruit on the edge of a pond, a thicket in a swale, a patch of cattails, a

Pitch pine, pen-and-ink botanical drawing, 1941.

swamp forest in a dune hollow, or a stand of birches on a stony hill slope all catch the eye. The dangers of personal reactions become evident when ecologists get involved in sophisticated abstractions. Most assume that the units that they describe function as communities, as if the world of each species is coextensive with what draws the ecologist's eye.

People are strongly drawn to create stereotypes, "typical forms," and then to match the variability in nature to them. For example, I can form a stereotype of "an oak leaf," even though the leaves in a single oak tree vary widely in shape and size, and the leaves of each species of oak tree differ remarkably. Between twelve and sixteen species of oaks are recognized in the northeastern states, excluding the live oaks, whose leaves "don't look like oak leaves." I can look over thirty or forty leaves and make a drawing that represents red oakness in a single leaf, in contrast with scarlet oakness or black oakness. I can, as a botanist friend said, "make them look more like a red oak than the real thing." It is all too easy to apply this process to vegetation, to form a mental picture of what is "typical" and stamp out vegetation units in the landscape.

If I list the plant species characteristic of one of these units, I find a few common and conspicuous species occurring together. These make up the "thickets" in swales, the "meadows" along a brook, the "piney woods" on sand plains, and the "heaths" on coarse soils. Botanists call these species dominant or primary species. Other less conspicuous species, called secondary species, occur with the conspicuous ones. The list of secondary species gets longer and longer the more ground I cover. The primary species determine my "units," even though inconspicuous species may be more numerous. This should make it clear that I can very easily become wedded to my own idea of a particular plant community and then ignore samples that—no matter how widespread—clash with my stereotype. If I add the fact that most of us accept as sophisticated the ability to recognize patterns that others miss, I have created a discipline that strongly resembles Impressionist painting.

Zones of vegetation on sandy beaches reflect patterns of wind strengths, wave length between dunes, and moisture levels. Sand spits are built out and plants migrate down them; then the spits are cut off or eroded away. Dunes form and are stabilized by marram grass; then blowouts start and move through the fields of dunes. The shifting sand creates and destroys islands where plants can live for a time. The geological processes impose comings and goings on plants and produce inconsistencies in how plant species combine into what people recognize as communities.

DEEPER QUESTIONS

Stress and Disturbance

Plant communities have been assumed to be organized around habitat differences, primarily the contrast between "mesic" and "stressed" sites.

"Mesic" is a vague term used for "normal" forested sites of temperate climates, while "stressed" is used for habitats too wet, too dry, too stony, or too salt-ridden. There is an assumption concealed in this: that plants prefer mesic sites but some are displaced to stressed sites by competition from other plants. It is as if botanists could not credit some species with being more comfortable in stressed sites—as Eskimos cope successfully in a habitat that would stress most of us.

Clements proposed that plant processes alter the conditions of stressed sites making them suitable for plants of more general adaptations (1916). This implies that he excluded the environment from the operation of the "system."

An alternative interpretation of vegetation is that "early successional" plants are adapted to locally stressed habitats. In other words, plants of low stature can tolerate blowing sand, drought, and salt spray. Taller trees and shrubs raise their sensitive growing points so high that they are at risk of wind and salt damage in the hollows behind the foredunes, but they can survive in sheltered dune hollows and overtop plants of lower stature there.

"Disturbance" also selects for plants of humble stature, and it deserves special comment because the idea implies that without disturbance the site would remain in the status quo ante. Pickett and White considered disturbance to be something outside of "the system": "A disturbance is any relatively discrete event in time that disrupts ecosystem, community, or population structure and changes resources, substrate availability, or the physical environment" (1985). John Wiens recognized that "because disturbance may be expressed on a variety of scales, it may produce extremely complex effects. It is little wonder that most theory ignores disturbance or considers it a random process" (1989). Most ecologists believed that disturbance was simply an interruption in a trend toward equilibrium.

These statements indicate a general agreement on an almost Newtonian concept for closed systems: a body at rest will remain at rest until "disturbed," and then "the body" tends to return to the original conditions. If communities are open, "stress" and "disturbance" are parts of "the system."

Simple Complexity

Most of what we have been taught as ecology has been about nature's complexity, and this has led us to put our faith in organizing mechanisms within natural entities. I argue that the invocation of self-organizing "communities" mystifies the natural world, suggesting that important elements in life cannot be understood. Here and in future chapters I offer alternative explanations that hold as closely as possible to the interactions going on in nature that you can see for yourself.

In the next chapter I will begin to move inland from the shore. I want to examine theories of interactions in a type of habitat that has been particularly popular in theories of communities: the salt marsh.

SOURCES

Bascom, W. 1980. *Waves and Beaches*. 2nd ed. Garden City, N.Y.: Anchor Books.

Clements, F. E. 1916. *Plant Succession: An Analysis of the Development of Vegetation*. Carnegie Inst. Washington Publ. No. 242:1–512.

Egler, F. E. 1977. *The Nature of Vegetation*. Bridgewater, Conn.: Aton Forest and the Connecticut Conservation Association.

Fox, W. 1983. *At the Sea's Edge*. New York: Prentice-Hall.

Lubchenko, J. 1978. Plant Species Diversity in a Marine Intertidal Community: Importance of Herbivore Food Preference and Algal Competitive Abilities. *American Naturalist* 112:23–39.

Lubchenko, J., and S. Gaines. 1981. A Unified Approach to Marine Plant-Herbivore Interactions. I. Populations and Communities. *Ann. Rev. Ecol. and Syst.* 12:405–37.

Lubchenko, J., and B. Menge. 1978. Community Development and Persistence in a Low Rocky Intertidal Zone. *Ecological Monographs* 59:67–94.

Menge, B. 1976. Organization of the New England Rocky Intertidal Community: Role of Predation, Competition, and Environmental Heterogeneity. *Ecological Monographs* 46:355–93.

Paine, R. T. 1963. Trophic Relations of 8 Sympatric Predatory Gastropods. *Ecology* 44:63–73.

———. 1966. Food Web Complexity and Species Diversity. *American Naturalist* 100:65–75.

———. 1969. A Note on Trophic Complexity and Community Stability. *American Naturalist* 103:91–93.

———. 1969. The *Pisaster-Tegula* Interaction: Prey Patches, Predator Food Preferences, and Intertidal Community Structure. *Ecology* 50 (6): 950–61.

———. 1971. A Short-Term Experimental Investigation of Resource Partitioning
in a New Zealand Rocky Intertidal Habitat. *Ecology* 52 (6): 1096–106.

Pickett, S. T. A., and P. S. White (eds.). 1985. *The Ecology of Natural Disturbance and
Patch Dynamics*. Boston: Academic Press.

Ranwell, D. S. 1972. *Ecology of Salt Marshes and Sand Dunes*. London: Chapman and
Hall.

Wiens, J. A. 1989. *The Ecology of Bird Communities*. Cambridge: Cambridge University Press.

Witman, J. D. 1985. Refuges, Biological Disturbance, and Rocky Subtidal Community Structure in New England. *Ecological Monographs* 55 (4): 421–45.

Woodwell, G. M. 1970. Effects of Pollution on the Structure and Physiology of
Ecosystems. *Science* 168:429–33.

3 Salt Marshes

I spent many hours as a teenager at a small patch of salt marsh at the mouth of a brook where beaches converged. When I went to college, I was taken as a member of the Birding Club to the impressive salt marshes and mudflats at the mouth of the Merrimack River. Salt marshes played an influential role in my early experience as a naturalist.

In my college course in ecology, salt marshes were treated as straightforward illustrations of the physiological effects of relative salt concentrations. It wasn't until later that salt marshes gained the limelight as a result of John Teal's ecological studies (1962, Teal and Teal 1969). Teal was a student of my general ecology professor, George Clarke. As a consequence of what these two started, ecologists know a great deal more about life in salt marshes and know it in a kind of detail lacking from study of almost all other ecological landforms. Attention has been directed to the marshes themselves and their growth, the species of animals that use the several parts of a marsh, the numbers and sizes of individuals of these species (the biomass), and the amount of energy that each represents. Hence we obtain a record of the biological material or energy that each set of organisms represents and how this matter and energy moves among the species.

THE IMPORTANCE OF SHELTER BEHIND THE BEACHES

Massive quantities of sand are washed away where storm waves beat against the loose sediments left by ice sheets. Sediments are carried by long-shore currents where velocity is high because flow is narrowed against a headland. The currents slow and drop their load as they pass the end of a spit that builds off the headland. This process can build a bar that closes off a river's estuary. Then the river that carries fine-grained material from inland drops its load in the shelter behind the bay-mouth bar, forming mud- and sand flats. Salt-tolerant grasses colonize the flats. Deposits of salt marsh peat build up around the grass.

From time immemorial, animals who have sought their livelihood at sea have gathered in havens behind barrier beaches. In the days of sailing and rowing boats, people built their fishing villages in coves behind sand spits like those at Provincetown Harbor, the mouth of the Pamet River, and Pleasant Bay in Massachusetts. The need for shelter is obvious.

Plant Species of a New England Salt Marsh

The coarse features of the pattern of the plant species in a salt marsh are distributed just as might be expected from theory. At the seaward edge grows a nearly pure stand of coarse cordgrass called midtide grass or thatch. Thatch grows on flats flooded during most of the tidal cycle and can colonize areas exposed only when the tide is two-thirds out. Another, finer cordgrass grows on the top of the marsh: this is high-tide grass or salt marsh hay. Salt marsh hay is short and characteristically forms fans of prostrate individuals spread like cowlicks.

Thatch can survive the longer submersion in seawater because it has salt-secreting glands on the surface of its stems and leaves. Equally important, it can survive the rapid sedimentation characteristic of the edges of the many deltas built into New England estuaries during the last thousands of years. Thatch's habitat in a rapidly sedimenting environment is like that of balsam poplar on the depositing slopes of the rivers of the north (see chapter 4). After the thatch colonizes, the individual stems and leaves interfere with the flow of the muddy water at the mouth of a river. Filamentous algae grow on the mud and the grass stems and trap sediments. Then as silt-bearing freshwater meets salt water, a physical interaction takes place that produces clumping in the fine sediments and hence rapid accumulation. Also for physical reasons, once silts and clays settle they cohere to the mud already there and resist being stirred into suspension. As sediments accrete, the level of the marsh rises. The thatch is conspicuous on the outer edges of the salt marsh, where it takes over shallow sand and mudflats, and along the banks of the many sinuous streams that drain the high marsh.

Salt marsh hay occupies the top of the marsh, which is wetted only at the highest tides. The border between the two conspicuous species of cordgrass may be broad in gently sloping areas but is characteristically narrow along the tidal creeks and their steep-walled gullies. The high marsh is richer in species than the intertidal areas.

Shallow pools, or pannes, usually lined with scumlike algae, are where emergent fleshy plants like samphire and aquatics like widgeon grass can be found. The levees along tidal creeks (where mud is dropped as floods subside) are vegetated with salt marsh hay, scattered clumps of spike grass and marsh elder bushes. On the still higher ground, along the inland margin of the salt marsh, the aspect changes rapidly. Large patches of black grass (actually a rush) become conspicuous.

The edges of the marsh are characteristically decorated with seaside goldenrod, sea lavender, and New York asters. A garland of mosquito-sheltering bayberry bushes can often be found up against the inner edges of the sand dunes.

The Animal Life of a Southern New England Salt Marsh

Nixon and Oviatt studied the kinds and numbers of organisms in a salt marsh in Rhode Island (1973). By counting and weighing individuals of the most numerous species, they could construct a picture of the local economy, an endeavor that delights the curious naturalist because it is another step in understanding a piece of the landscape. Salt marshes are highly productive of organic material. Teal, working along the Savannah River in Georgia, found that productivity depended on thatch grass (1962). Nixon and Oviatt also found that most of this "primary production" came from emergent thatch, although in the salt marsh pannes widgeon grass and the algae and diatoms encrusting the sediments made major contributions. In other places, green algae such as sea lettuce and filamentous forms that grow in the drainage channels and shallow pools in the mudflats, contribute by providing raw materials for fiddler crabs, marsh mussels, marsh snails, sandworms, and grass shrimp. In addition, the grass shrimp chew the thatch, breaking it into small bits and making it available as detritus, decaying material that bacteria colonize.

Detritus-feeders take in the particles of indigestible plant bits, digest the bacteria, and spit the "chaff" out. Many different individuals and species will process the detritus as bacteria repeatedly colonize, grow, and are eaten. Not surprisingly, a number of organisms come in from the shelf waters beyond the mouth of the inlet to take advantage of the available nutrients during parts of their lives when they grow or fatten up prior to breeding or migration.

As tides flood in through the inlets, up the channels, and into the salt marsh creeks, bait fish like mummichogs follow, searching for the worms, grass shrimp, fiddler crabs, and other prey feeding on the algae or the organic detritus. The bait fish and crabs are followed by larger predators: sea bass, bluefish, common terns, red-breasted mergansers, green herons, black-crowned night herons, and snowy egrets. The pattern is what ecologists have called a food chain.

TRADITIONAL SYSTEMS ECOLOGY

In developing ideas of ecology, it was easy to conclude that the thatch growing where the tidal flow is deep—at the advancing edge of the expanding marsh—prepares the way for the establishment of the salt marsh hay, and that the increasing species diversity on the high marsh is related to the maturity of the system. During the 1930s it became accepted that the organic elements that ecologists called communities were integrated into the inorganic, forming a more all-encompassing system. The influential British ecologist Arthur Tansley coined the term *ecosystem*: "In an ecosystem the organisms and the inorganic factors alike are components which are in relatively stable dynamic equilibrium. Succession and development are instances of the universal processes tending towards the creation of such equilibrated systems" (1935). We can see here how the classical model of succession and climax played a central role in the thinking of a new field of physiological ecology.

Energy Flow in Ecosystem Function

The trophic-dynamic model, originated by Lindeman, establishes a special level of organization, implying that the members of a community react among themselves while the community as a whole reacts with the physical environment (1942). Green plants capture energy from the sun and are eaten by herbivores who are eaten by carnivores who may be eaten by second-level carnivores. Eventually, of course, all individuals die and are decomposed by bacteria and fungi, and the products of decomposition are reused by the green plants. The system recycles its nutrients while energy flows from the sun through the system. Energy flow and nutrient cycling are the glue that holds communities together.

The idea of energy flow is an attractive one for teaching and useful in categorizing data. It has obvious parallels to money flow in a human economy. Money, its movement and manipulation, seems to make industrialized societies work. But while the amount and kind of energy available sets limits on the kinds and numbers of organisms in a system, energy moving along the lines of predator/prey interactions does not perform an organizing function or "drive the system." Humans are different in that they can make a reality out of an abstraction. An institution in a human society can control the money supply and cash flow. But, whatever the reality, the existence or nonexistence of such ordering institutions in nature has constituted one of the major controversies in ecology.

According to the theory of succession, which I will be examining later in some detail, as additional species enter the system and thereby increase the number of functions being performed, the complex of community functions is improved. Eventually, it is argued, a stable climax community is formed, and it persists because it is well integrated and maintains more efficient physiological functions than do successional stages. So the "normalization" of both structural and functional relations drives the community. As Lindeman put it: "From the trophic-dynamic viewpoint, succession is the process of development in an ecosystem, brought about primarily by the effects of the organisms on the environment and upon each other, towards a relatively stable condition of equilibrium" (1942).

A novel intellectual framework like that provided by the trophic-dynamic model stimulates a different way of looking at things, and trying any different perspective is likely to provide interesting insights. But this ordering system has so many exceptions and contradictions that I wonder at its value as a general model. Energy flow seems to have little relevance to the lives of individual animals.

The tightly linked world of the climax community is by no means the only interpretation that can be placed on the life of a salt marsh. The narrow creeks provide access to the debris that comes out of the grasses at the margins of the marsh. Much of the energy may pass from bacteria to worms to flounder to cormorants, eels, or ospreys, or from bacteria on suspended organic particles to mussels to eiders or scoters.

The mussels, clams, flounder, sea bass, codfish, old squaws, scoters, eiders, harbor seals, and commercial fishermen may seem to us to dominate the scene, but none are necessary for an entirely satisfactory operation of the "system"; the "higher forms" are extras, fifth and sixth wheels. They are physiologically unnecessary embellishments of the system. Remove the embellishments and the remnants will readjust, but the "system of survivors" will still "work." What the evident or edible (to us) organisms do not eat will decay, or, we might say, pass through the bacterial food chain. The chemical elements, nitrogen, potassium, phosphorus, carbon, and oxygen, will be recycled. Even in the total absence of the "higher" organisms, the green plants and the bacteria or fungi can function effectively and with equanimity.

The Pyramid of Numbers and
Concentration of Persistent Pesticides

One of Charles Elton's helpful rules of thumb was the observation that the relation between predators and their prey takes on the form of a pyramid of numbers (1927). The number of predators is very much lower, about 10 percent of the number of the prey population. Hence, as Colinveaux put it, "big, fierce animals are rare" (1978). A consequence of this rule was vividly demonstrated when members of the chlorinated hydrocarbons (DDT, chlordane, heptachlor, dieldrin) were found to become concentrated in the fat of predators at the tops of food chains.

One of the advantages claimed for these chemicals when they came into general use after World War II was that they "stayed on the job," that is to say, they persisted in the environment. But this meant that they remained in the fat deposits of animals such as invertebrates, fish, frogs, and snakes that lived in the bodies of water where the polluted runoff from agricultural land collected. They also persisted in the bodies of predators that ate the insect pests, or those that fed where agricultural runoff collected in estuaries.

The chlorinated hydrocarbons were carried in organic litter or on soil particles by runoff into the mud of small streams and down rivers to mudflats where bloodworms and tube worms processed the sediments and were in turn fed on by flounder. Osprey fed on the flounder. The ospreys received an increasing pesticide load with each flounder they ate because the chemicals were concentrated at each level of the food chain. When female ospreys came to lay eggs, they drew on their fat reserves, releasing the DDE they had concentrated in those reserves. The DDE interfered with an enzyme system involved in laying down calcium in their eggshells; the eggshells of contaminated females were 20 percent thinner than normal and readily broke.

Many predators were affected: bald eagles, ospreys, peregrine falcons, black-crowned night herons, red-shouldered hawks, and Cooper's hawks. As the use of the persistent pesticides increased, the populations of the vulnerable species declined and some, like the eastern population of peregrines, disappeared. After many vigorous programs collecting massive amounts of evidence, and especially vigorous and well-organized lobbying programs at local, state, and federal levels, the use of these pesticides was restricted, then banned, in the United States.

As concentrations of the chemicals in the soil and in the lower levels of consumers declined, the input was removed and the system purged itself. The surviving populations of the vulnerable birds began to recover. Ospreys and bald eagles continued to fail for a number of years in parts of their ranges, but some adults survived and now the populations have rebounded.

This evidence of movement of a chemical through a system is a metaphor for movement of energy through trophic levels, from its being "fixed" by green plants until it is lost as heat by a warm-blooded apex carnivore. Many ecologists consider this to be a major driving force of ecological systems.

At the time of the pesticide controversy, many vocal ecologists argued that apex predators are critical in the functioning of entire ecosystems, that they tie the system together by regulating the numbers of prey. This was also a time when widely quoted ecologists were making the seductive argument that species diversity is critical to the health of ecosystems. Diversity, it was argued, leads to stability by adding connections in food webs. Intuitive support for this argument came from the observation that prudent money managers generally maintain a diverse portfolio of investments. It was believed that the fewer the connections in simple systems, like those of the far north or in industrialized agricultural systems, the less stable the overall system would be. It seemed evident that when species diversity decreases, a community is destabilized. This in its turn provided dire foreboding for the destabilizing effects of drastic reductions in species diversity by deforestation of the wet tropical jungles.

AN ALTERNATIVE VIEW:
THE IMPORTANCE OF EXTERNAL INFLUENCES

Rise of Sea Level and the History of a Salt Marsh

Alfred Redfield's findings in his study of a salt marsh on Cape Cod showed that the salt marsh had not always been as extensive or as thick

Bulrush, pen-and-ink botanical drawing, 1941.

as it was then (1965, 1972). In fact, peat right on top of the glacial sediments at the mouths of creeks contained freshwater species: cattails, bul-
rushes, and reeds.

One hypothesis for the distribution of thatch and salt marsh hay based
on field observations would be that because the thatch forms a garland
on the outer margins of the expanding areas of marsh, most of the peat
should be thatch, with a frosting of salt marsh hay on top. This would be
consistent with the theory of succession: that sediments and organic remains collecting around the thatch prepare the way for salt marsh hay to
colonize, once the marsh level has reached high tide.

The many holes that Redfield dug showed the opposite: that the peat
consisted almost entirely of salt marsh hay, all the way to the bottom.
Thatch peat occurred only on the outer edges of each layer of peat, as if
growing on the rim of a pie plate, just as thatch is distributed on the
marshes today. How could this be?

The answer is that silt is deposited by river floods that cover the top of
the entire marsh. The increasing segments of the marsh are small-scale
deltas over whose entire surfaces sediments collect. Sedimentation keeps
the tops of the deltas at high-tide level. The rise of sea level allows the salt
marsh hay to occupy the top of the marsh all the time, and allows sedimentation to continue to fill upward and expand the salt marsh surface
seaward. As the addition of sediments builds up the marsh, the tidal
edge, where thatch grows, moves away from the uplands.

Meanwhile thatch continues to provide garlands on the seaward edge
of each growing delta. Thus, salt marsh hay peat occurs as a wedge over
the thatch peat, decreasing in thickness toward the outer margins of the
marsh. Thatch peat also appears in a wedge that reflects the gradual rise
of sea level, which creates the threshold below which thatch doesn't grow.
Salt marsh hay peat also extends in a wedge into the uplands over the sediments laid down in freshwater before the incursion of salt water.

At the same time that the sand spit at the mouth of the marsh has been
building out to sea, gradually increasing the area sheltered from storm
waves, stream deposition has deposited sediments farther out in the inlet. The combination creates a larger and larger area of the inlet that of-

fers the right combination of sediments, water level, and shelter from waves for cordgrasses to establish. Both thatch and the salt marsh hay plants colonize the mud and sand surface as fast as their particular surface reaches above the critical level relative to low tide. The combination allows the salt marsh to grow outward from the heads of the creeks onto the flats. So the inlet becomes filled by a broad, marsh-covered delta.

The delta is not simply a local phenomenon. Its form is influenced by the circulation of ocean water that leads to the movement of sand along the beaches. Moreover the sediments that have filled the inlet have been carried downstream in the rivers that empty into the inlet. Thus to an important degree the high productivity of the estuary depends on silts carried off the uplands in soil erosion somewhere inland. The conditions that converge on the salt marshes find their origins at some distance.

How Individuals Crop the Resources in Estuaries

Now I want to emphasize the openness and ephemerous nature of interspecific associations.

The strong currents that sweep past the sand spits beside tidal inlets act as endless belts, carrying food past the would-be diner. By standing and watching, a lazy observer can save a lot of energy and yet search the equivalent of many miles of water. Gulls, seals, and people—all omnivorous opportunists—converge where turbulent waters and shifting tides expose fish and shellfish, and where currents carry dead things past with the tide. Gulls epitomize the users of this habitat. Their tastes are cosmopolitan. They are looking for virtually anything edible: shrimp, cockles, mussels, scallops, crabs, starfish, sea urchins, sea worms, small fish, carcasses, beetles, young terns and ducks, discarded hot dog rolls, and the heads, guts, and backs of fish discarded by fisherfolk. This active habitat of shifting tides, weather, and food supplies is appropriate to what is perhaps gulls' most important adaptation—the ability to compensate for change. They may do this better than any other seabird, and their wide geographic distribution reflects their tolerance of a wide range of differences.

A concept like energy flow credits natural food webs with a degree of organization focused on the whole. It would be more useful to consider the excess of young produced under the goading of natural selection that makes trophic levels possible. I will develop this theme in later chapters: parents raise as many young as they can in their lifetimes or they would be replaced by others that did. The number produced is much larger than the number of young that can become established breeders. This gross excess is the "stuff" that makes herbivory and carnivory work.

How do waterfowl use the marshes and lagoons? Are they residents settling to become closely associated with a particular salt marsh or estuary, or are they nomadic sojourners gleaning what they can and moving on? In Rhode Island, Nixon and Oviatt found that the numbers of birds on the marsh were lowest in the fall when the numbers of their presumed prey were highest (1973). Active vertebrates are mostly nomadic during the winter, gathering where they find a concentration of food, cleaning it out, and then leaving.

The estuarine world is the epitome of an open system. Nutrients come to it from soil erosion far inland, carried by rivers. The residents of estuaries —seaweeds, blue mussels, quahogs, bay scallops, spider and horseshoe crabs, moon snails, channeled whelks and sand fleas—release clouds of eggs and sperm into the flood tide. Their larvae are carried in the plankton miles up and down the beaches. When the larvae settle, chance takes a terrible toll, but the wastage has its advantages. Populations eliminated by scouring of storm waves or smothered by an oil spill can be replaced the next spring by clouds of new young.

Young of several species of marine fish, such as smelt, flounder, silversides, and menhaden, grow from larvae to small fish in estuaries. Mature individuals of other species such as winter flounder, eels, striped bass, and bluefish come into bays to feed, often on the growing bait fish, worms, shrimp, scud, and crabs that have been feeding on detritus. Other marine fish, such as salmon, alewives, and shad, pass through estuaries on their way to spawning grounds upriver. The young of these come downstream again to take up their life in salt water. Transient flocks of waterfowl suddenly appear, feed on the excess, and are gone.

During spring migration in May and fall migration from July through

September, sandpipers and plovers gather to feed furiously at low tide on worms, snails, scud, and small bivalves at a few special places where broad mudflats enjoy or suffer a bloom of concentrated food. These shorebirds lay on the fat that fuels their thousands of miles of overland or overseas flights, in some cases from the coasts of New England nonstop to Argentina.

Movements between alternate feeding grounds and the use of different foods and shifting tactics seem to be characteristics of estuary feeders. Evans and Dugan reported that sanderlings at the Tees Estuary in northeast England were almost sedentary, while those feeding along the coast patrolled many tens of miles, visiting a number of beaches (1984). The roving bands were checking the quality of potential feeding areas on beaches subject to removal and redeposition by violent storms and tides.

When prey is mobile and feeding areas made unpredictable by physical factors, birds are not likely to show good correlations between their density and prey density, except over very short distances and times. Instead they prudently hedge their bets by spreading out and moving. Under these conditions, it is hard to see how equilibrium conditions or the notion of some fixed carrying capacity of the mudflats could apply.

The general picture that I have presented is epitomized by the multiplicity and variability of lagoons inside the barrier beaches that line the Alaska coast. A barrier beach extends for thousands of miles along the northwest coast of Alaska, built by storms and long-shore currents. The beaches hold in small bodies of water, each of which has its own special characteristics. As I flew low over these beaches counting gulls and waterfowl, it became clear that the birds aggregated into certain lagoons, leaving many unoccupied. A series of accidents of depth and recent history influences the amount of food and attractiveness of the lagoons. As long as redundancy is provided by the thousands of coastal lagoons, then at least some will always provide just the right, rare combination of conditions that makes a lagoon productive and attractive. As long as birds can follow the shore until they find one that is just right, chance is enough to supply them habitat. But, if the number of lagoons is reduced by development, then chance will not suffice, and it may be necessary to manage some stretches of shore to ensure the proper habitat conditions.

SOURCES

Colinveaux, P. 1978. *Why Big Fierce Animals Are Rare.* Princeton: Princeton University Press.

Elton, C. 1927. *Animal Ecology.* London: Sidgwick and Jackson.

Evans, P. R., and P. J. Dugan. 1984. Coastal Birds: Numbers in Relation to Food Resources. In *Coastal Waders and Wildfowl in Winter,* edited by P. R. Evans, J. D. Goss-Custard, and W. G. Hale, 8–28. New York: Cambridge University Press.

Lindeman, R. L. 1942. The Trophic-Dynamic Aspect of Ecology. *Ecology* 23:399–418.

Nixon, S. W., and C. A. Oviatt. 1973. Ecology of a New England Salt Marsh. *Ecological Monographs* 43:463–98.

Redfield, A. C. 1965. Ontogeny of a Salt Marsh. *Science* 147:50–55.

———. 1972. Development of a New England Salt Marsh. *Ecological Monographs* 42 (2): 201–37.

Tansley, A. G. 1935. The Use and Abuse of Vegetational Concepts and Terms. *Ecology* 16 (3): 284–307.

Teal, J. M. 1962. Energy Flow in the Salt Marsh Ecosystem of Georgia. *Ecology* 43 (4): 614–24.

Teal, J. M., and M. Teal. 1969. *Life and Death of the Salt Marsh.* New York: Ballantine Books.

4 Rivers, Floodplains, and Bogs

I now want to move upstream, away from the coast, examining the fact that many of the ideas that we have already encountered have been applied to other regions. I shall also introduce the roles that differing schools of geology have played in shaping our notions of ecology.

When you look out over a landscape of mountains, valleys, and broad plains, you are drawn to think that streams erode their way into the hills at their headwaters. It seems reasonable to believe that the present lower reaches were once in the hills, and that these hills have now been washed away. In other words, the differences evident now in space represent changes over time. One finds echoes of this changing landscape in the Bible: "Every valley shall be exalted, and every mountain and hill shall be made low, and the crooked shall be made straight, and the rough places plain" (Isaiah 40:4). This logic is familiar. If one adopts a closed-system view of the world, one is likely to accept that both the process and the products of erosion must start somewhere and go somewhere else, with the river *as a whole* constituting the "system." This is akin to a metaphor of embryological development for both landscapes and vegetation. From an open-system perspective, each length of a river is expected to respond to local conditions, not some long-term developmental goals.

I am going to focus on meandering rivers and the forests and wetlands on their floodplains. The theme of the chapter is deteriorating order. We start with seemingly easily quantifiable concepts that on closer examination reveal local inconsistencies. The physical systems involved, like the form of the river—its width, depth, velocity, and the sweep of its meanders —exhibit an admirable orderliness and consistency. The regularities that the river imposes by cutting short the lives of some forests and burying the floors of others with sediment sets a stage for regularities in the forests that botanists have interpreted to be inherent in the vegetation.

While we might speculate about the lengths of time available for plant succession on the uplands, by studying the history of river meanders we can establish and measure the relations between time and events in riparian areas. It is a place where we can test whether changes in vegetation follow the predictions of succession.

HOW A RIVER WORKS

A river takes on a three-part form all through its existence: headwater streams, middle reaches, and lower reaches. This is the way a river works:

On the upper reaches of a river, the streambeds are steep and valleys narrow. Where the bed is steeper, the flow is stronger and the river carries fine-grained materials and small gravel, leaving behind coarser stones. Periodic violent floods in the headwater brooks carry away all but the large stones found in the streambed.

Where rivers come out of the hills, their beds are still fairly steep and they drop coarse debris as they emerge onto a broad piedmont slope. The gravels are moved in occasional big floods, and they dam and deflect the river at low water flow. The fans of debris work their way up into the lower parts of the valleys.

On the broad lowlands outside the mountain front, meandering main streams flow through fine-grained materials. Water contributed from many tributaries maintains a relatively constant discharge; the river flows in a meandering channel on a gentle slope. The stream cuts steep cutbanks in the silty floodplain and carries a load of silt in suspension.

In its lower reaches the river is running close to base level and meanders widely across a broad floodplain. Its muddy load is carried suspended in the turbulent flow. The main channel is deep and narrow and its banks are steep. The river runs on deep sediments left when previous floodwaters subsided. These fine-grained sediments have been carried off the uplands in mud slides. The coarse sediments too heavy to carry are left behind.

The point where a river empties into a major body of water, the base level of the drainage basin, sets a point at which there is no slope, and hence no erosion. All upstream channels are keyed to the base level at the lower end of the reach, as well as to the amount and rate of delivery of water and sediments from above. For example, an outcropping of bedrock can create a dam and slow the flow, or engineers can straighten a channel and steepen the bed and speed the flow. What happens along each stretch affects the lower end of the next stretch upstream. Thus, streams in rocky hills alternate between splashing rapids over steep, rocky beds with quiet pools over pebble bottoms, and meandering snakelike flows between sandy banks along a gently sloping stretch above the next outcrop of bedrock. The stream returns to its headlong plunge as it reaches the ledge, and it shifts as quickly back to meandering where it rushes out of the hills onto salt marshes in a rocky cove. The stream's form is determined as much by the rate of flow out of the bottom as by what is put in at the top. This is why no single mathematical formula can account for the longitudinal profile of even a short river.

Each reach of the river is involved in erosion, transportation of the sediments, and deposition where slope or water volume falls below a threshold. If too much sediment or too-coarse particles are delivered to the stream, deposition occurs. This steepens the bed and increases the water's speed until sediments are moved again. Wherever a stream crosses a rocky outcrop its bed is steep, its velocity increases, and its stony load actively chips away at the bed.

THE FORMATION OF RIVER MEANDERS

One of the impressive things about rivers is that their romantic beauty can be represented both in an artist's aesthetically pleasing lines and an engineer's rigid mathematical formulations. As you steer a boat up a broad river you become aware of the regular swing of the meanders and the regular shifts of the channel back and forth across the river. Day after day, the straight stretches become shorter, the river narrower, and the turns tighter. The channel runs next to the cut-bank on each bend, crosses over in the straight stretch, and hugs the cut-bank on the next bend.

Most country children know a brook contains steep places (riffles) alternating with pools (reaches). These shift from side to side up and down the stream bed. The back-and-forth movement seems to result from the speed of the water as it plunges over the ground. When its speed becomes too fast the water "goes around," the way a piece of cooked spaghetti you try to push across a plate takes on a sinuous form. Rivers flowing over the ground or over ice, and even the streams of water flowing on the surface of the ocean, take on this form, turning to one side and then the other. This is because water can only flow so fast down a straight channel due to the effects of cohesion of water molecules (viscosity) and bottom friction (shear).

This sinuous flow causes the stream to erode its banks at points regularly spaced along the channel. As the jet is confined by running against the outside of the bend, it flows faster and cuts against the banks on the outside of the curve. The river running against the bank sets up eddies, which concentrate energy and focus the erosion. As erosion proceeds, bends appear and the sediments carried off a cut-bank settle out where the water slows on the crossover bar or on the point bar on the inside of the next bend.

The floodplain of a river slopes downhill, and gravity acting along this slope draws the stream to turn downslope, counteracting the tendency to turn very far in any one direction. The process of erosion and deposition continues until the longitudinal profile of the stream is adjusted to discharge the energy available. Empirical evidence from field studies shows a strong correlation between mean annual discharge and meander length, between channel width and meander wavelength, and between channel width and the radius of curvature of the meander.

The waveform approaches the curve that defines minimum total work of turning. A curve generated by the sine function best represents the shape of river meanders. Any curve other than a sine-generated curve would, by increasing the angle at which the river turns, tend to concentrate bank erosion at certain places and would add to the total erosion and hence "work."

The river eats its way across the floodplain as it cuts on the outside of the meanders, and deposits on the inside of the bend. Erosion of the concave banks and deposition on the convex banks make the meanders move sideways across the river valley. Because the process is random, the combined migration of meanders over many years results in the river channel eventually occupying all possible positions between the valley walls. Deposition from the flood-borne sediments, combined with successive occupation of all parts of the valley floor, builds the familiar broad, flat river floodplain.

EFFECTS ON VEGETATION

The changes in landscape wrought by the shifting pattern of a river have profound effects on the vegetation of the floodplain. The patterns produced by geological and hydrological processes are carried over at least in part to the biological realm. Unfortunately, many biologists have been ignorant of or have chosen to ignore the full impact of geology and hydrology on theories of vegetation. In what follows I will endeavor to show the importance of understanding geology, using examples from my own work and that of others in the far north.

My time in Alaska as a student of Hugh Raup exposed me to 750 miles of the channel of the Kuskokwim River and one of its major tributaries that drained out of the Alaska Range (Drury 1956). The zones of forest trees on the floodplains of meandering streams have been used to illustrate vegetational succession, and the forests of the Kuskokwim fit readily into the standard sigmoid curve describing an initially gradual and then more rapid increase in stature and complexity as the vegetation achieves its maximum development. In most cases botanists have ignored geological processes that are going on while the forest stands are changing, apparently assuming that the surface remains unaltered once the silts of a floodplain are deposited. The stated assumption is that the differences in vegetation reflect progressive vegetational development. Several studies made of the forests on floodplains of rivers in northwestern America shed light on how the case is different when looked at closely.

The zone of active meanders is readily recognized by the linear traces of abandoned river channels in the forests of Alaska. Troy Pewe proposed two terrain units for this zone most frequently visited by the river in its writhing (1948). Pewe's first terrain unit consists of the growing slip-off slopes inside the bends. Here the curving lines of recently deposited bars and swales and recently abandoned channels are distinct and parallel to the present river. The higher ground is occupied by deciduous species: willows and alders on the point bars and tall poplars on the slip-off slope behind them. The second terrain unit is older (seventy-five to one hundred years) and is in the area from which the river has moved away. Here the patterns of bars, swamps, and oxbow lakes have been blurred, and they lie at an angle to the present river. The higher ground is occupied by homogeneous stands of tall coniferous trees.

The most active part of the floodplain is repeatedly covered by muddy floodwaters that do not reach the older areas. The floodwaters create a natural levee higher than the bottomlands against the valley sides. Thus the higher ground supports vegetation different from that on the older parts of the floodplain for simple geological reasons. But botanists have generally misidentified the open-floored, mixed forest of tall white spruce and white birch as the climax forest.

The geological phenomena involved in the continuous building and destroying of a floodplain surface sets a certain life expectancy for the surfaces. The deposition sets the stage on which plants live their lives, and the rate at which meanders move downstream limits the time that the stage will be available. The admirable orderliness of the forms of the river deteriorates as secondary geological effects and plant growth smudge the clean lines found in the zone of active meanders.

In the Zone of Active Meanders

In a meandering river, a continuous slope exists from the deep water off the cut-bank up onto the point bar and along the long axis to the peninsula created by the advancing meander loop. The outer, newest parts of the point are lowest and subject to the heaviest rates of deposition. According to traditional succession theory, the roots of grasses and willows stabilize newly deposited sediments and contribute organic material that facilitates the establishment of the taller longer-lived trees, which, being more shade tolerant, are believed to have greater competitive abilities.

The zones of vegetation described by Nanson and Beach for the slip-off slope of the Beatton River in British Columbia are typical of the vegetation of active meanders of the rivers of northern Canada and Alaska (1977). Willows, balsam poplar, and mountain alder are the primary woody colonists on new surfaces, and white spruce is dominant on older surfaces. Nanson and Beach counted annual rings at the bases of trees growing on the floodplain and used the results to date events. Their counts of tree rings suggest that the zones of vegetation on the slip-off slope are responsive to the rate and amount of silt deposited by the river in flood rather than to the age of the surface, even though the two geological processes are closely related.

Annual deposition of sediments adds a new ridge to the outer edge of the depositing slope every 27 years on average (between 5 and 75 years depending on the size of floods). Vegetation establishes on these surfaces as soon as sediment accumulates above the summer low-water level. Yet, "after new floodplain surfaces are high enough above the channel to support tree species, an additional 2.5 m of sediment may be added to both ridges and swales over a 50-year period" (Nanson and Beach 1977).

Trees are progressively older the farther they are from the river. A major point that Nanson and Beach made is that different trees grow in different zones, depending on the trees' abilities to tolerate different frequencies of flooding and amount of siltation. Rapid sedimentation favors the establishment of dense, nonreproducing stands of balsam poplars and continuous regeneration of willows and mountain alder.

The density of balsam poplar increases for the first 30 years of ridge formation, reaching its maximum at 40 to 50 years. The rate of sedimentation declines after this time, by which time deposition has built the level up to a height that is above most floods. Within 10 years, successful white spruce seedlings colonize ridges under the poplar canopy. Sedimentation continues at a much slower rate between 50 and 250 years, and after 250 years sedimentation contributes little change to the floodplain surface.

Balsam poplars are less frequent on surfaces older than 80 to 100 years and absent from surfaces more than 200 years old. The decrease of balsam poplar releases white spruce seedlings that have become established. Spruces are maximally dense on ridges between 100 and 200 years old and are the only trees present on ridges 200 to 300 years old, forming even-aged, nonreproducing stands in their turn.

Traces of geological change can be found on the oldest parts of the floodplain, changes that occurred in the distant past and continue down to the present. They show that geological activity occurs so frequently that the vegetation exists in a constant state of flux.

Beyond the Zone of Active Meanders

In Alaska the natural settling of sediments away from the zone of active meanders has spectacular effects. Once sedimentation declines, woodland and *Sphagnum* mosses colonize and carpet the shaded forest floor; white spruce declines and is replaced by black spruce. The ground freezes in winter and is insulated from thawing by a dry moss mat in summer. This stimulates lenses of ice to segregate in the silty soils. As the older parts of the floodplain settle, the water table rises, flooding and allowing the expansion of bogs into low places in the forest. The details of bog topography and combinations of plant species vary as much within each bog and from bog to bog as they do from river basin to river basin.

Some of the lowlands are occupied by bogs simply because the water table rises. While the low ground is being taken over by *Sphagnum*, species that grow when submerged occupy the margins of still waters, filling the interstices of a loose mat of horsetails and sedges. The moss and sedge clones fan out in tan, brown, and light and dark green scallops through the water, held together by runners of river horsetail, tall beaked sedge, and some rare wildflowers. As the water table rises, the sedge meadows expand their domain, obliterating the outlines of oxbow lakes.

In other parts of the lowlands, the patterns are different, reflecting the fact that the floodplain silts become frozen. As bog-trotters know, the edges of the bog are very loosely vegetated along their shores. The water in this moat is warmed by the sun, and, where it laps against the frozen silt in the bank, thaws, undermines, and causes slumping in the bank. As the banks slump, the bog expands.

The oldest parts of some floodplains have another pattern. On these surfaces, which usually lie on recognizable terraces, lakes of irregular shapes and scalloped borders occur in large numbers, with the islands of "high ground" consisting of stunted muskeg. The ground is frozen and ground ice is conspicuous. The lakes and bogs on the surface occupy depressions formed by thawing of the frozen ground. But, the universality of permafrost and ground ice suggests that this pattern is a remnant of an earlier, colder climate.

At present the pattern of lakes and bogs is in flux. Slumping in cut banks expands the area of "thaw sinks," and these perpetuate their own growth. Thaw sinks may become shallow ponds whose shores retreat as summer winds blow warm water against the lee shore (Black and Barksdale 1949). Loosely woven mats of sedges and *Sphagnum* colonize newly formed ponds. The ice, blown by spring winds as it melts away from the shore, is pushed against the loose sedge meadows, creating triangular sets of peat ridges and outlining the ghost of what was a thaw lake.

The variation of patterns in different places suggests that in some of those places warm temperatures favor thawing, and that thaw sinks are expanding over the lowland. But in others the bog mosses and sedges are growing actively enough to colonize shallow waters and clog the lakes.

Walker sampled bog sediments in Britain from the surface down to the gravels left by streams flowing off the ice ten thousand years ago (1975). This allowed him to trace the identity of plants growing together since the retreat of the ice. The evidence showed no orderly change in any "direction."

In summary, the implications of the Darwinian model to which I subscribe are that the patterns of vegetation that we see in the landscape today reflect the local conditions of site and contemporary aspects of slope, soil, and physiographic processes. Each topographic unit is believed to provide peculiar characteristics for plant habitats: ridges, hillside slopes, landslides, coves, terrace edges, and floodplains. The number of species native to the region and topographic type determines whether each site has a vegetation type specific to it. When influences change, new and probably different assemblages appear swiftly in response to new local conditions. Any combination of assembled species "works."

SOURCES

Black, R. F., and W. L. Barksdale. 1949. Oriented Lakes of Northern Alaska. *Geol. Journal* 57:105–118.

Drury W. H. 1956. *Bog Flats and Physiographic Processes in the Upper Kuskokwim River Region, Alaska.* Contrib. Gray Herb. No. 178.

Dury, G. H. 1970. General Theory of Meandering Valleys and Underfit Streams. In *Rivers and River Terraces,* edited by G. H. Dury, 264–75. New York: Praeger Publishers.

Hack, J. T. 1960. Interpretation of Erosional Topography in Humid Temperate Regions. *Amer. Journ. Sci.* 258A:80–97.

Leopold, L. B. 1974. *Water, a Primer.* San Francisco: W. H. Freeman.

Leopold, L. B., and W. B. Langbein. 1966. River Meanders. *Scientific American* (June): 60–70.

Leopold, L. B., and T. Maddox. 1953. *The Hydraulic Geometry of Stream Channels and Some Physiographic Implications.* U.S. Geol. Survey Prof. Paper 252. Washington, D.C.: USGPO.

Leopold, L. B., M. G. Wolman, and J. P. Miller. 1964. *Fluvial Processes in Geomorphology.* San Francisco: W. H. Freeman.

Nanson, G. C., and H. F. Beach. 1977. Forest Succession and Sedimentation on a Meandering-River Floodplain, Northeast British Columbia, Canada. *Journal of Biogeography* 4:229–51.

Pewe, T. 1948. *Terrain and Permafrost on the Falena Air Base, Galena, Alaska.* U.S.G.S. Permafrost Progress Report No. 7.

Viereck, L. A. 1970. Forest Succession and Soil development Adjacent to the Chena River in Interior Alaska. *Arctic and Alpine Research* 2 (1): 1–26.

Walker, D. 1975. Direction and Rate in Some British Post-Glacial Hydroseres. In *Studies on the Vegetation History of the British Isles,* edited by D. Walker and R. G. West, 117–39. Cambridge: Cambridge University Press.

Whittaker, R. H. 1975. *Communities and Ecosystems.* 2nd ed. New York: Macmillan.

5 Geological Activity and the Response of Vegetation on Upland Slopes

In the earlier chapters, I identified the power and pervasiveness of geological processes along the seashore and riverbanks, where their effects are obvious. In this chapter I will show that similar although less evident processes are active in the uplands, breaking stones, churning sediments, and moving earth downslope. These processes are active at many levels of intensity and frequency as they mold the form and content of what geologists call unconsolidated surficial deposits. Because of this multiplicity of levels these processes strongly influence the distribution and abundance of plants.

THE MARK OF ICE ON THE LAND

Now I shift to what are recent events on a geological time scale that, however remote they are to us, still influence plant distributions. Continental

67

Rufous-sided towhee, pen and ink, circa 1965.

ice sheets left the mark of their heavy hand on the landscape of the north-eastern United States and Canada. The ice polished rock surfaces on the shoulders of hills and scoured out valleys, cutting off the lower reaches of tributaries. Like a giant bulldozer it pushed cubic miles of surficial un-consolidated material out of northern uplands to the edges of the low-lands. The ice left morainic tills and outwash at its edges and piled up the fishing banks from Newfoundland to Cape Cod. It seems impossible to exaggerate the extent and importance of the ice's impact, but its very ex-istence has been acknowledged for barely twice my lifetime.

Glacial Scour

Ice alters topography by gouging out preglacial valleys. It grinds hard against the sides and bottoms of valleys, plucks angular rock fragments off bedrock, fixes them in its grasp, and uses them as tools to grind other outcrops as it moves over them.

Matthes described ice scour on the granitic mountains of Yosemite Na-tional Park (1950). Similar effects, on a less grand scale, are evident in Aca-dia National Park in Maine. Massive granite is resistant and ice moves up-slope over it, polishing the upslope surfaces. On the lee side, the ice freezes onto exposed blocks and carries them away. Hence, granitic moun-tains show spectacular, angular cliff faces and rounded domes.

More friable bedrock (metamorphic and sedimentary rocks) is shattered by frost action. The frost rives peaks and ridges that stand out above the ice and heaves the rubble downslope. The riving of broken rocks carves jagged peaks and sharp ridges. Mountains of such friable materials show isolated jagged peaks and great sweeping lines of broken rock on convex slopes. The rapid erosion of exposed peaks and ridges reinforces the tendency of mountain tops in glaciated regions to be at similar levels.

The ice under the glacier presses heavily as it moves, deepening and altering valleys from V-shaped (convex slopes) to U-shaped (concave) with sheer valley sides.

The Drift

Rock debris left by the ice or the streams flowing off it is known as drift. The great bulk of the drift was removed from northern uplands and pushed to the exposed coastal shelf between the ice and the deep sea. Ice came and went many times, and traces of different ice sheets are evident as loose, younger till and weathered-compacted older tills. Some remnants were plastered on valley sides and in valley bottoms under the ice. These are ground moraines. Some were moved to the edges of the ice where the ice stood temporarily. These make up lateral and end moraines.

Glacial deposits tell you how they were formed, more than what they were beforehand. Till, the muddled mixture of coarse stones, sand, and stuff as fine as clay, is a major component of glacial deposits. Compacted tills packed into lower slopes are usually partially covered by material that slumped off the valley sides, creating the gentle "run-out" of lower slopes in glaciated topography.

The ice may override debris, pushing up rounded, elongate hills that look like overturned boats, called drumlins. Drumlin fields are scattered along the coasts of Nova Scotia, around Boston, and in Wisconsin, where preglacial topography apparently jarred the flow of ice. Other topographic signs of glacial landscape include undulating topography of round hills and closed valleys, small lakes, and sand ridges. The landscapes along the glacial border from Minnesota and Iowa to Pennsylvania and Long Island are barely subdued versions of those in the north, where the ice is today a few miles away. Moraines left by earlier advances are muted, as frost heaving and slump have moved materials off the ridges into low places.

The prairie potholes, west and southwest of the Great Lakes, are examples of ground moraines. The low rolling hills are pockmarked with undrained hollows. Where underlaid by tills, the hollows have "perched" water tables, making for multitudes of lakes, ponds, and marshes. Rainwater percolates through lenses and sheets of sand extending for miles under the till, and it emerges as artesian springs that maintain the water table in the lakes and marshes. These ephemeral wetlands make high-grade nesting grounds for waterfowl and waders. One pothole may attract half a dozen species of nesting ducks and several sandpipers.

When the last ice melted away from the coast twelve thousand to thirteen thousand years ago, the land was still pressed down by the weight of the ice, and the ocean flooded far up the hillsides and into New England valleys. Breaking waves built cobble beaches two hundred feet above the present sea level on granite ridges on Mount Desert Island and along glacial moraines fifteen miles from the sea farther east. At this time, rivers flowed turbid with clay and silt, which settled out, plastering much of midcoast Maine with marine clays containing the shells of blue mussels and Arctic species of clams. As the coast rebounded, the sea retreated, but it still fills the lower reaches of preglacial valleys. The rocky ridges separating valleys are now peninsulas and the hilltops are islands.

Frost heaving in the ground around the margins of the ice sheets churned the upper levels of the soil, eliminating the orderly zones of leached and deposited minerals that we associate with soil profiles. British farmers noticed this and called it the warp. New England soils show evidence of churning associated with the intense frost activity. Indeed, frost action created many of the fine-grained topographic features on New England uplands. Most people are unaware of these features' existence, let alone of their pervasive influence on the distribution of trees.

The vegetation of the mountains responds to the effects of solifluction, or frost-moved rubble, and vertical movement of fines on patterned ground, because of the effects the soil structures have on the distribution of moisture. Once your eye is trained to see these responses you begin to see similar ones lower down. The forest trees there also respond to the effects the soil structures have on the distribution of moisture.

VEGETATION RESPONDS

The shape and size of a small valley reflects the importance of the size of a drainage basin. First-order streams occupy much smaller drainage basins than do second- or higher-order streams. The importance lies in the frequency of intense, localized downpours of rain in the hills. A single heavy rainstorm may drench the entire drainage basin. Such downpours, like spring freshets that follow a sudden thaw, produce spectacular local floods. The effective forces carving valley forms are periodic downpours, landslides, and floods—all rare events of unusual intensity.

Between downpours, fallen trees, branches, and leaf litter collect on slopes and in channelways. Long-term studies of a forest stand in the drainage basin of a small headwater stream, Hubbard's Brook in the White Mountain National Forest in New Hampshire, have clarified what is happening in the physiology of the forest (Bormann and Likens 1979). Teams of botanists and chemists followed the chemical interactions between the trees, the litter, the soil organisms, and the mineral soil. For decades, organic materials and mineral nutrients cycle between organic debris in the soil and the wood in trees. Leaf and branch litter collects in valley bottoms and may block the stream channel. High winds add branches and occasionally whole trees to the debris on the forest floor. Following fires or cutting, the trees grow back and production of organic matter exceeds respiration for many years. Then botanical processes slow as the canopy is closed. Although, for short-lived botanists, the forest seems to be "at equilibrium," sooner or later geological time takes over. A downpour carried on especially high winds knocks the forest down and the floods move the debris out of the creek beds. During these special events sediments start their travels to the sea.

John Hack, a geomorphologist, and John Goodlett, a forest ecologist, worked together in the headwaters of the Shenandoah Valley, in the heart of the Appalachians, in the 1950s (Hack and Goodlett 1960). Hack and

Goodlett's study was inspired by one such spectacular flood on the Little River in Virginia. They found massive new gravel and sand deposits all up and down the river, both in the alluvium and in tongues of debris on the valley sides. The landslides of surficial debris loosened from valley sides and left scars of exposed mineral soil. Equally important, they found conspicuous evidence of previous landslides. By dating the trees on the old debris fans, they showed that major slides had occurred well before the arrival of European colonists.

Hack found several generations of debris slides extending back several hundred years, and Goodlett found several species of forest tree, such as scarlet oak, to be closely associated with the coarse soils at the tips of these "fossilized" avalanche tracks. The chutes created by landslides, twenty to one hundred feet across, were most numerous in the hollows at the heads of valleys over outcrops of Hampshire shale and sandstone, a weak rock. The landslides began just below ridge crests, under outcrops of Pocono sandstone, a resistant rock.

The evident relation between landslides and vegetation had confused earlier foresters who interpreted the correlation as reflecting the effects of logging in the late nineteenth century. It was logical within classical American ecological theory to consider the second growth, which grew up following the logging, to be immature.

To attribute the flood damage to the "inferiority" of the vegetation carries the implication that stable mature vegetation would have prevented it. But Hack and Goodlett pointed to debris from many landslides older than the logging that had been colonized by the same species that grow on the recent debris fans. The debris from the older landslides showed that the suggestion that an undisturbed forest can protect the soil from landslides leads to a contradiction. The two researchers observed that "compound debris fans are unequivocal evidence of flood damage to slopes in the past, prior to the period of logging and probably prior to the time of settlement by the white man. Either the forests mantling the slopes in the past were not all 'thrifty normal stands'—a possibility—or the slope forests, regardless of quality, cannot control huge volumes of runoff" (1960).

Furthermore, if the vegetation is developing toward a regional norm, then progressively older landslides should be vegetated by species closer to the regional norm. But Goodlett found the species on the oldest land-slides to be the same as among the seedlings on the newest slides.

Hack and Goodlett found the distribution of forest trees to correspond closely to topographic divisions and contemporary geological processes. Goodlett found that altitude, exposure of the slopes, soil moisture, and kind of bedrock strongly influence the distribution of forest trees. He found that species distributions correlated best with features that affected the availability of water through the season. Details of local conditions were especially important. He found that the moisture-loving and fibrous-rooted yellow birch grew on rubbles along streams and on block fields where snow and ice persist, and that the drought-tolerant tap-rooted oaks may be excluded from the hollows by frost heaving.

FORESTS OF ONCE−FROST-ACTIVE SOILS

The continental ice reached the central uplands in Pennsylvania. Goodlett surveyed the forests on residual soils beyond the glaciated area and on the drift within the glacial boundary to see what might be the effects of dif-ferences in history (1954). You would expect, a priori, to find stark differ-ences. South of the reach of the ice, you would expect deep residual soils, developed in situ during long periods of chemical weathering. In contrast, where the ice had reached, you would expect the soil to have been scraped off ridge tops, leaving blocky deposits in pockets behind rock shoulders and on valley bottoms.

As he expected, Goodlett found thin till on the hilltops and long, smooth, boulder-strewn slopes in the glaciated areas. But instead of the expected marked differences between the glaciated and residual soils, he found a thin, disturbed, and structureless soil spread over the entire land-scape, under the current organic layer. The structureless layer included fine-grained material studded with large pebbles, boulders, and frag-ments and a large amount of fine sand and silt, which appeared as if wind-blown. Many of the large rock fragments were rounded and derived from the underlying bedrock, while others were angular, derived from bedrock further upslope.

Goodlett found no differences in the tree species composition over a wide range of topography on the two sides of the border, but he did find important differences in species according to present climate, topography, and soil structure. While the general vertical distribution of forests reflects climate, local differences reflect the composition of surficial deposits, especially the abundance of rock fragments that affect the depth to the water table.

In her important book on the forests of the Appalachians, Lucy Braun suggested that the oak-chestnut forests occur on younger surfaces of the Appalachians because the soils on those surfaces are too young to allow the full development of the climax mixed mesophytic forest (1950). Extending the logic to the south, she believed that the mixed mesophytic forests persisted since the Tertiary on undisturbed soils on the highlands of the Cumberland Plateau. Extending the logic to the north, she suggested that the forests within the region of the Wisconsin Drift are immature because the soils have had less than ten thousand years to mature since the retreat of the last ice.

Cooper's studies in Glacier Bay, Alaska (where glaciers have been retreating rapidly during the last one hundred years), are used to illustrate the principle that pioneer mat plants and then shrubs such as willows and nitrogen-fixing alders are required to prepare raw soils for colonization by trees (1923). The principle is weakened if trees colonize raw mineral soil on glacial moraines within a few years of being uncovered by glacial ice. Sigafoos and Hendricks examined the time intervals between stabilization of alpine glacial deposits and establishment of tree seedlings at Mount Rainier, Washington. By counting tree rings, they found that "where tree ages were compared with known ages of recently deposited moraines . . . the interval between sediment stabilization and tree seed germination is estimated to be 5 years for bottomlands" (1969).

Why then is it so far from the ice tongues in Glacier Bay to the forests far down the valley? That is still a good question, but perhaps local winds and frost action discourage establishment farther up the valleys. Sigafoos and Hendricks found that "dating of a recently deposited pumice layer and determining the ages of trees growing on an underlying moraine near timberline show that the interval is longer at high altitudes than at lower altitudes" (1969).

Moreover, M. B. Davis reported that in the postglacial sequence of vegetation reported from pollen analysis of deposits in peat bogs, "an interesting aspect of the migration of spruce is that everywhere spruce preceded alder (*Alnus*), the pollen of which peaked *after* spruce began to decline. . . . This is somewhat surprising, given the well-known primary succession sequence at Glacier Bay, where nitrogen fixation by alder is important in allowing spruce to grow" (1980).

I doubt that an orderly succession involving alders played an important role in the establishment of a white spruce forest sixty to seventy feet tall perched on two feet of morainic debris over green-blue glacial ice that I found on the Matanuska Glacier, north of Anchorage, Alaska. I found no alders in the understory.

Field studies have shown that the seedlings of many trees survive best on mineral soil. The young do not need a mature soil to become established, and they create a new profile faster than they were given credit for. Olson reexamined chronological details of soil development in the Indiana sand dunes studied by Cowles (Olson 1958). He reported: "Soil analyses of carbon, nitrogen, moisture equivalent, carbonates, acidity, and cation exchange relations show how most soil improvement of the original barren dune sand occurs within about a thousand years after stabilization. The pattern of change on older dunes promises little further improvement and perhaps even deterioration of fertility" (1958). According to this, 12,500 years is more than enough time for vegetation to create a soil profile.

VEGETATION ON GLACIAL AND FROST-MOVED SOILS

In the morainic topography along the coast of Maine, many kettle holes and depressions have been taken over by bogs. The bogs are typically domed, their tops determined by the height to which dead stems, leaves,

Plate 3. Bylot Island, Northwest Territories, Canada. The distinct ridges in front of the tents were formed by frost action. Photo by the author.

and *Sphagnum* peat can lift water by capillary action. The capillary action must be especially effective toward the middle of the bog for the bog to be domed. In many bogs, low places toward the center "degenerate" into wet meadows and ponds that form a concentric pattern around the dome. It would seem that the hydraulic processes exceed the botanical and that water drowns places on the top of the dome. Under such circumstances, it is imprudent to consider these bogs to be temporary stages of vegetational development in ancestral ponds, being filled with organic debris and destined to be overgrown by forest.

Not surprisingly, the clearest correlations between vegetation and glacial deposits are found where glacial topography is best defined. One such place is the ground moraine of the prairie potholes from Nebraska and Iowa through Minnesota, the Dakotas into Manitoba and Saskatchewan. The rain stands in the depressions, underlaid by compacted tills while it quickly drains away from the rounded sand hills. The topography and soil conditions have clear and dominant influence, and they differ over short distances. The depressions are occupied by aquatic plants while the higher ground is occupied by grasslands. The borders are sharp between the aquatic vegetation and grasslands. It appears to be more than coincidence that Clements gained values-forming experience about consistency and naturalness of plant associations in these places.

During most of the twentieth century, academics have considered the typical habitat of white pines to be a place in the succession of old fields, but the pine was abundant and it called a different place home long before the white settlers came. Most of the pines grew on outwash plains left by rivers flowing off the stagnating glaciers. Nearly pure stands covered the extensive sand plains of southern New Hampshire and central and eastern Maine. I shall discuss the importance of the growth patterns of white pines in chapter 7.

DEEPER QUESTIONS

In this chapter I have focused on geological processes in surficial materials on hill slopes. Even though the movements of material may not be demonstrable in the short attention span of a human, they have effects on the growth of trees on slopes and they leave patterns in coarse and fine-grained materials that favor some species here and others there.

While chance is uppermost, much of the variance in local distribution of forest trees is explainable in terms of the complex of soil texture, soil moisture, and topographic exposure that foresters refer to as site. This is the botanical equivalent of zoologists' "niche." Appreciation of these concepts shifts attention away from equilibrium communities as the units of study to questions about the distribution of individual species, alternative combinations of species, and how plant associations (and the concept of plant associations) vary from place to place.

The consistency we see in plant associations must be influenced by our ability to "keep track" of a number of species at the same time. As long as people restrict their experience to a small region (the experience of a local naturalist) they can keep in mind a list of species associations that are consistent within their home range. While I was pursuing the study of plant communities, I found I could pick out patches that matched what I had in mind, but when I looked closely and kept track, I realized I was steadily expanding the list of infrequent species.

Gleason and later Preston realized that a small number of species are abundant in most samples of vegetation (Gleason 1926; Preston 1962); a few more occur in small numbers in most samples; but most species are infrequent and scattered, in fact relatively "rare." Gleason described this frequency distribution as an exponential decay curve. This leads to the question: how are variations in species associations distributed over major geographic areas?

My strong impression is that the recognition of consistent plant communities is largely limited to mid latitudes. The phenomenon results from a regression of highly variable local plant lists against a nearly constant range of variation in microtopography, exposures, and soils. Differences in microtopography, and so on exist almost independent of latitude and position on continents, but local climates vary widely according to latitude and local conditions and have dramatic effects on the richness of the flora. Places with "moderate"—warm and moist—climates are relatively rich in species, while those with marked swings in environmental conditions support only a few.

The coniferous forests consist of a small number of species relative to the number of species found in warm temperate forests or the tropics, but a large number of species relative to tundra regions. "Plant associations" are woven out of shifting combinations of these species, certain combinations occurring repeatedly because the species are common and widespread. Beyond that, chance plays a central role.

When the number of species present matches the topographic diversity, an observant naturalist can match consistent clusters of species to recognizable habitats. We should not be surprised that the concept of communities developed where the number of species is "not too small, not too large, but just right," as Goldilocks might have said, to match local microtopography. This happy coincidence exists where there are moderate floras (in central and western Europe and the eastern Great Plains of the United States). It just happens that in these locations schools of plant sociology developed, closely associated with major midwestern universities. In contrast, botanists who have worked either in the north or in the tropics have been less sanguine about community concepts.

SOURCES

Bormann, F. H., and G. E. Likens. 1979. *Pattern and Process in a Forested Ecosystem.* New York: Springer-Verlag.

Braun, L. 1950. *Deciduous Forests of Eastern North America.* Philadelphia: Blakiston.

Cooper, W. S. 1923. The Recent Ecological History of Glacier Bay, Alaska. Parts I, II, and III. *Ecology* 4 (2): 93–128; (3): 223–46; (4): 355–65.

Davis, M. B. 1980. Quaternary History and the Stability of Forest Communities. In *Forest Succession,* edited by D. C. West, H. H. Shugart, and D. B. Botkin, 132–53. New York: Springer-Verlag.

Gilluly, J., A. C. Waters, and A. O. Woodford. 1975. *Principles of Geology.* 4th ed. San Francisco: W. H. Freeman.

Gleason, H. A. 1926. The Individualistic Concept of the Plant Association. *Bulletin of the Torrey Botanical Club* 53:7–26.

Goodlett, J. C. 1954. *Vegetation Adjacent to the Border of the Wisconsin Drift in Potter County Pennsylvania.* Harvard Forest Bulletin No. 25.

Hack, J. T. 1960. Interpretation of Erosional Topography in Humid Temperate Regions. *Amer. Journ. Sci.* 258A:80–97.

Hack, J. T., and J. C. Goodlett. 1960. *Geomorphology and Forest Ecology of a Mountain Region on the Central Appalachians.* U.S. Geol. Survey Prof. Paper 347. Washington, D.C.: USGPO.

Leopold, L. B., M. G. Wolman, and J. P. Miller. 1964. *Fluvial Processes in Geomorphology.* San Francisco: W. H. Freeman.

Matthes, F. E. 1950. *The Incomparable Valley.* Berkeley: University of California Press.

Olson, J. S. 1958. Rates of Succession and Soil Changes on Southern Lake Michigan Sand Dunes. *Botanical Gazette* 119 (3): 125–70.

Preston, F. W. 1962. The Canonical Distribution of Commonness and Rarity. Part I. *Ecology* 43:185–215, 431–32.

Prince, R. J. 1973. *Glacial and Fluvioglacial Landforms.* New York: Hafner.

Sigafoos, R. S., and E. L. Hendricks. 1969. *The Time Interval between Stabilization of Alpine Glacial Deposits and Establishment of Tree Seedlings.* U.S. Geol. Survey Prof. Paper 650-B. Washington, D.C.: USGPO.

6 Contrasting Perceptions of the Landscape and Forests of the Appalachians

In this chapter I will introduce two contrasting perceptions of landscapes and vegetation. The members of one major school of thought accepted landscape change as progressive toward an equilibrium, while the members of the other school considered landscape change to be open-ended, responding only to local conditions. The two schools emerged at the turn of the century. Both were available to geographers, but the equilibrium school was nearly universally accepted while the other was largely ignored.

I will begin by reviewing the model of landscape "development" that William Morris Davis proposed for the Appalachian Mountains (1899). I will show how the model related to work done by two leading proponents of the idea of succession, Cowles and Clements, and how it influenced the botanist Braun's interpretation of the history of the major forest types in the Appalachians. Then I will describe how Margaret Bryan Davis's 1980 study of the pollen record of post-ice-age changes in the ranges of tree species presented a very different picture.

THE APPEARANCE OF LANDSCAPE SEEN AS A DISTANT PROSPECT

Almost all geographical studies from the first fifty years of this century were cast in a mold that assumed that long-term geological processes reduce once-rugged topography to a level plain. This plain then persists for millions of years until uplifted and caught up in a new cycle of erosion. These geological assumptions influenced botanists to propose long-term plant successions proceeding to a mature climax vegetation that remains until some sort of catastrophe intervenes. A lot of animal ecology has in turn been based on assumptions that mature vegetation provides homogenous and stable habitats. The importance of this chain of reasoning cannot be emphasized too much, as Chorley said: "The interpretation of a given body of information depends as much upon the character of the model adopted as upon any inherent quality of the data itself" (1962).

Lacking the benefit of the theory of plate tectonics (which was not introduced until 1915 and not accepted until the late 1960s), W. M. Davis in 1899 proposed a theory of cycles of erosion in which periods of intense uplift occurred as a block of continent "bobbed up." The continental mass then remained free of uplift for tens of millions of years, during which time rivers eroded the uplifted mass from the continent borders towards the center. As the mountain front retreated, it left behind broad lowlands where aged streams meandered across a nearly level plain, referred to as the "peneplain," which remained until the entire landscape was again caught in a period of uplift.

W. M. Davis reasoned that the last major cycle of erosion in the Appalachians had started with the mountain building that created the New England Alps in the Permian Revolution. By the end of the Cretaceous cycle of erosion, the Appalachian region had been reduced to a peneplain, across which rivers flowed wherever they happened to meander. Renewed uplift at the start of the next cycle during the Cenozoic stimulated rapid erosion and oversteepened streambeds. The courses of main rivers however, were already set in the channels they followed on the Cretaceous surface, hence as they eroded they cut "entrenched meanders" on the topography below.

W. M. Davis portrayed the major features of overall landscape change as they might be seen from a distant prospect. From the top of the Blue Ridge you can see the flat tops of mountains to the west. According to W. M. Davis, despite difference in the sedimentary rocks that outcrop at the summits (he called the effect "discordance of structure"), these flat topped mountains at similar altitudes ("accordance of summits") are the remnants of the uplifted, partially dissected Cretaceous Peneplain, which he called the Schooley Surface. He identified another erosion surface, the Harrisburg Partial Peneplain, on the high shoulders of many valleys, and considered it to be the product of a later and even less complete cycle of erosion. He credited the development of the Harrisburg surface to erosion during the millions of years between the mountain-building at the end of the Mesozoic Era (that uplifted the Schooley Surface) and renewed mountain building in the Miocene period, about 30 million years ago. Moreover, he acknowledged that the Schooley Surface was not perfected in the region of the Cumberland Plateau in Tennessee. These remnants of mountains older than the Schooley Surface persisted because they were remote from the river systems that carved the Schooley Surface.

This view of the geological history of the region had a great effect on botanical interpretations of both the Appalachians and other areas. As Chorley put it: "[W. M.] Davis' approach to landforms was to study them as an assemblage, in which the various parts might be related in an areal and a time sense, such that different systems might be compared, and the same system followed through its sequence of time changes" (1962).

Cowles and Clements clearly felt that the basic unit of the study of vegetation is the plant community (Cowles 1899, 1901; Clements 1916): an assemblage in which the parts (species) can be related in space and time and that can be followed for changes through time. They used the word "succession" to imply progressive replacement of communities, step-by-step preparation of a site for the vegetation of the next stage, and achievement of a predetermined end point controlled by climate. This end point is the "climax," which they suggested persists until destroyed by some form of disturbance. The climax is directly analogous to W. M. Davis's peneplain. Whittaker identified what he called "trends or progressive developments [that] apply to most successions" (1975). These included progressive development of the soil; increased stratification of the plant community; increased stock of inorganic nutrients held in plant tissues; increased productivity; community determination of local microclimate; increased species diversity; change in populations over time in a manner similar to that over space; reduced replacement rate; and increased stability. All of these "trends or progressive developments" are important basic assumptions that underlie conventional ecology. We have encountered them before, and we will again.

Clements and many botanists after him concluded that because the forest or prairie climax is consistent over a region, it must remain continuous through time. As members of this school argued, species within a mature community may react among themselves, but the community as a whole reacts with the physical environment (Clements and Shelford 1939; Odum 1969, 1983). This encouraged them to believe that where remnants of the Schooley Surface survived unaltered during the ice ages, they might find the survivors of ancient floras.

In his widely quoted paper, Fernald proposed that the vicarious species he found on Newfoundland mountain tops, which presumably had not been overtopped by the Wisconsin ice advances, were relics of pre-ice flora that had persisted on the Tertiary surfaces that protruded above the continental ice (1925). Although Fernald's idea has been discredited since then, it influenced many young botanists. It encouraged them to believe that land surfaces had been sculptured during the Tertiary and survived unaltered since then. Consistent with this thinking, botanists concluded that the woody flowering plants that increased and expanded during the Tertiary—after the great extinctions at the end of the Cretaceous—persisted on unaltered peneplain remnants.

According to this viewpoint, as millions of years passed the woody forests of the Tertiary became increasingly restricted to areas of diverse topography, especially as a late Miocene uplift began a new cycle, stimulating erosion among the weak rocks and carving the great Appalachian

Valley. Paleobotanists believed that this partial erosion cycle eliminated the Tertiary forests from lowlands and isolated less-fully-developed forest types on a partial peneplain surface on the shoulders of the large valleys.

Traditionalists suggested that climate changes of the Pleistocene were not severe enough beyond the glacial boundary to displace the mixed mesophytic forests from the presumed deep soils of the Cumberland Plateau. These climax communities would have resisted invasion while northern forests retreating before the ice were able to infiltrate the coastal areas where the shallow soils of the piedmont were unsuitable for the species of the ancient forest. At the same time, opportunistic, subclimax oak forests expanded over the Harrisburg remnants.

The Forests of the Appalachians

Lucy Braun made use of her access to the mountains as a Tennessee volunteer nurse in the 1930s to prepare a monumental work on the old growth forests. She published a summary article (in 1947) and an important book (in 1950) that was the definitive description of the forest landscape of the region as perceived within W. M. Davis's and Clements's developmental model.

In her book, Braun clearly stated her assumption of long periods of stable landscapes where soils and forests develop:

> Here also, physiographic history seems to have had a profound effect upon the nature of present-day vegetation. On the ravine slopes formed in the latest erosion cycle, the vegetation is developing in response to present forces both topographic and climatic. Although very limited in extent, it seems logical to assume that mixed mesophytic forest is the potential climax of the area; that by its development the extent of the mixed forest, greatly restricted at one or more times during the Tertiary and early Pleistocene, may expand eastward into what we now know as the Oak-Chestnut region. The outliers are forerunners in a development which would take thousands of years to complete, for it must await the development of land surfaces and of soils no longer related to the Harrisburg cycle. That these mixed mesophytic communities are not relics of a former more extensive mixed mesophytic region (mixed forest of the Tertiary) seems certain be-

cause of their limitations to surfaces produced in the last (or present) erosion cycle. (1950)

Braun's interpretation of the deciduous forests was built on the assumption of regional succession. She believed that the forests richest in species, the mixed mesophytic forests of the Cumberland Plateau in Tennessee, were remnants of the fully mature regional climax forest. She believed that this forest had originally developed on the Schooley Surface and had persisted on the remnants of that peneplain during the ice advances.

Braun also believed that the forest communities of eastern North America had migrated south during the climate deterioration that coincided with the Pleistocene ice advances and were gradually regaining their previous geographic distribution and mature structure. She classified the less diverse forests on lower slopes (oak-hickory) and on the mountains farther east and north (oak-chestnut) as being less developed successionally. She concluded that the oak-hickory forests would eventually develop to climax, and that they were in their current condition because of the lack of time since the retreat of the ice.

She assigned great importance to the separation of the deep residual soils south of the glacial margins from the raw, recently exposed sediments north of the ice margin. The direct connection between the W. M. Davisian theory of landscape development and Clements's theory of vegetation is in the soils.

In W. M. Davis's scheme, as the erosion cycle starts, bedrock outcrops, waterfalls, and rapids are common, while deep soils persist on the remnants of the older surfaces. As the cycle proceeds, soils thicken and develop, especially downstream, so that at the end of the cycle the whole lowland peneplain is covered with deep erosional and residual deposits. The classical scheme of soil development assumes that mature soil types are associated with mature vegetation types, both being products of a long period of growth and integration under constant environmental conditions.

The mature soils and the climax vegetation occur on the deep alluvial or residual deposits of the mature land surface—the peneplain. Each climax and soil type is determined by the regional climate. Local effects of drainage, topography, and bedrock are temporary in the development of soils and the peneplain. It seems clear that Braun thought through her classification of the forests of the central and southern Appalachians so that their development matched this story of landscape development.

AN ALTERNATIVE VIEW: THE FORESTS AS SEEN FROM THE GROUND

We have already encountered some of Hack and Goodlett's work in the previous chapter, but it is important to further examine their studies in contrast to the W. M. Davis–Clements-Braun interpretation of the landscape.

The Distribution of Forest Types by Topography

Goodlett perceived the forests in terms of the exposure and angle of slope and of the current geological processes. These effects are mediated by soil types, particle size, and availability of moisture. While Braun assumed that deep soils on the flat tops of the ridges supported the oldest and most mature forests, Goodlett found the ridge tops to be occupied by early successional pine forests on shallow soils. Braun assumed that the highest development of the local forests—the northern hardwood forests—would be on the most stable soils. Goodlett found the northern hardwoods to be on relatively unstable soils in the coves at the heads of tiny streams subject to mud slides in periodic downpours.

Braun assumed that the most mature forests on the old peneplain surfaces would contain the highest level of species diversity. Goodlett found that the bottomlands of third-order streams have floodplains showing exceptional species diversity. This increased diversity was maintained by the conditions created by floods and landslides. The seeds of upland trees are carried off the slopes during rainstorms and they germinate in the debris fans among the regular residents.

Braun relegated oak forests and chestnut forests to an incomplete erosion surface, the Harrisburg surface. Goodlett found the oak forests on the valley side slopes, where contour lines are evenly spaced and slightly convex. Furthermore, oaks were common under outcrops of the Pocono sandstone, the resistant rock that maintained the ridge tops in this part of Virginia. The taproots seem to give oaks an advantage in reaching moisture deep in the rubbles on side slopes.

Hack found that profiles along the ridge crests and convex upper slopes followed a simple exponential relation between the horizontal distance from the ridge top, the vertical distance below the crest, and constants related to the slope's steepness and curvature, whatever the bedrock. In other words, the shapes of the upper slopes are products of contemporary erosional processes. He found no support for, or need to call on, the assumption that the upper mountain slopes are remains of an ancient peneplain. W. M. Davis's original suggestion that accordance of summits existed with discordance of structure would require that Hack's measurements be dismissed as coincidences.

Several geologists have shown a close relation between the altitude of the ridge tops and the sediments in the outcrops there. For example, the higher mountains are all underlaid by the most resistant rock in the area, Pocono sandstone. Hack also showed that the gradients of the Harrisburg Partial Peneplain can be explained by present drainage patterns without invoking partial dissection of a previous surface (1965).

If we compare the developmental models of landscape and vegetation, it becomes evident that Hack's and Goodlett's observations contradict major predictions of W. M. Davis, Clements, and Braun. The summits are not "discordant in structure." All can be correlated to strata of the resistant Pocono sandstone. The form of the upper slopes conforms to a generally applicable rule reflecting the movement of materials, so the suggestion of remnants of an older peneplain is irrelevant. The "climax," or most mesophytic forest, is found on unstable slopes, not on the hilltops presumed to be remnants of a peneplain. Maximum species diversity—most closely approaching the theoretical climax—is found on the repeatedly disturbed bottomlands, not on stable upland sites.

These observations directly contradict the idea of succession from xerophytic early stages through late successional stages toward a stable climax. The fact that the occurrence of tree species on each of the topographic types can be explained by the features that affect soil moisture belies the idea of development gradually emancipating the vegetation from local environmental conditions.

STABILITY AND MOVEMENT
OF FOREST COMMUNITIES

Braun assumed that the climax forest that she had projected on W. M. Davis's peneplain was a stable structure that would have persisted intact in the face of repeated ice advances unless actually overwhelmed by catastrophe. This persistence would have been both a result and a cause of the self-organization of the climax community. Under Clements's ideas of succession a great deal of time would be needed to achieve the climax, but once it had been achieved it would last relatively unchanged indefinitely. Margaret Bryan Davis showed that Braun felt that the climax forest would have retreated as a unit as the ice advanced, and then migrated north again as the ice retreated: "Before data were available documenting the nature of vegetation south of the ice sheet, paleoecologists assumed that climatic changes associated with growth of glaciers displaced existing forests, as intact communities, southward" (1980).

Have the members of plant associations been constant over time or are they being extensively reshuffled with changes in climates? Good evidence about the conditions of climate and vegetation in the past comes from fossilized pollen grains preserved in peat deposits. M. B. Davis contributed to our knowledge of post-Pleistocene migrations of forest trees in the eastern United States. She summarized studies of pollen taken from deposits all over the area east of the Rocky Mountains: "The migration routes for individual species differed greatly; some moved from east to west, others moved from west to east. Rates of movement also differed. . . . During much of the Holocene, however, the distribution of many species were in disequilibrium with climate. . . . The speed with which forests can adjust to climate is thus far exceeded by the speed of climatic change" (1980).

Be reminded that, as a consequence of the most recent changes in climate, tree seedlings that became established in New York in the 1960s enjoyed climatic conditions similar to those experienced by seedlings that became established in North Carolina during the Civil War. And, it appears, species of forest trees have been crowded together in many different combinations in the past, then dispersed and pressed together again, presumably aggregating into different combinations with each reduction or expansion in range. M. B. Davis observed: "During the Quaternary Period the forests with which we are familiar seldom maintained a constant species composition for more than 2000 or 3000 years at a time. The evidence . . . suggests that forest communities in temperate regions are chance combinations of species without an evolutionary history" (1980).

This means that neighbor species can be expected to change over space and time. A "plant association" or "plant community" consists of a "collection of strangers" whose membership is defined by recent historical accidents, chance effects in seedling dispersal, and the ways in which colonists match the characteristics of the place. It is important to recognize that this view of the world would have come as no surprise to Gleason: "It may be said that every species of plant is a law unto itself, the distribution of which in space depends upon its individual peculiarities of migration and environmental requirements. Its disseminules migrate everywhere, and grow wherever they find favorable conditions. The species disappears from areas where the environment is no longer endurable. It grows in company with any other species of similar environmental requirements, irrespective of their normal associational affiliations" (1926).

The contrast in perceptions that we have seen in this chapter seem to reflect the degree to which the authors think of the landscape, vegetation, and other organisms as an entity that persists over time and that takes on its own identity. Many people accept the idea of a species persisting through time, and for several decades many ecologists applied the same concepts to patches of vegetation that they recognized as entities. These ecologists assumed that their entities persisted as "species of vegetation," that is, distinct community types. This meant that they assumed that the unit was to an important degree free of influence from surrounding forces, that it was a system, one rather loosely connected with its environment.

The differences between the two physiographic schools (those who subscribe to W. M. Davis's model and those who subscribe to that used by Hack) were identified by the geographer Chorley as reflecting "closed" versus "open" system thinking (Chorley 1962). In the simplest terms, closed systems differ from open ones in that a closed system has a border and its energy and materials are injected into it at the start of its processes. During the course of change (which is deterministic—headed toward a predictable end point), the potential energy is dissipated, materials are altered, and the system achieves equilibrium at its end point when energy is no longer effective in altering materials. This is ecology viewed as a lake. In contrast, in an open system, energy and materials are constantly moving through, and the system's form and processes are maintained by the movement of energy and materials. This is the ecological model of the river; the system has no predetermined equilibrium. A river will run as long as the rains fall, and the rising curve that describes a streambed persists as long as rivers run. It will take on the form that reflects the kind and amount of water energy and sediment matter brought to it; that is, it is indeterminate or probabilistic.

While geologists, spurred in part by increased knowledge of plate tectonics and the return to respectability of the idea of continental drift, have moved away from the deterministic pattern of W. M. Davis's theories, ecologists have been slower to adjust their thinking. In many cases they have adopted an even more stringent degree of closure for organisms in the environment; for example, consider the rigid structures of systems ecology. The abstract concept of the community or the ecosystem becomes in a sense more "real" than actual organisms.

Enormous amounts of effort are invested in studying and managing ecosystems, even though the practitioners involved will usually confess when pressed that they cannot identify the boundaries or even the full composition of their "object" of study. Underlying much of this work is a basic assumption that in the absence of humans, wilderness will itself evolve to produce a balanced harmony of best use, defined in terms of some set of tangibles such as primary productivity, biomass, or species diversity.

It seems to me that those who conclude that a wilderness climax consists of the best of all possible worlds subscribe to closed systems models. Those who see nature as rather shoddily arranged, poorly planned, and therefore potentially benefiting by enlightened intervention, subscribe to open system models.

SOURCES

Braun, L. 1947. Development of the Deciduous Forests of Eastern North America. *Ecological Monographs* 17 (2): 213–19.

———. 1950. *Deciduous Forests of Eastern North America*. Philadelphia: Blakiston.

Chorley, R. J. 1962. *Geomorphology and General Systems Theory*. U.S. Geol. Survey Prof. Paper 500-B. Washington, D.C.: USGPO.

Clements, F. E. 1916. *Plant Succession: An Analysis of the Development of Vegetation*. Carnegie Inst. Washington Publ. No. 242:1–512.

Clements, F. E., and V. E. Shelford. 1939. *Bio-Ecology*. New York: John Wiley and Sons.

Continents Adrift and Continents Aground: Readings from Scientific American. 1970. With introductions by J. T. Wilson. San Francisco: W. H. Freeman.

Cowles, H. C. 1899. The Ecological Relations of the Vegetation on the Sand Dunes of Lake Michigan. *Botanical Gazette* 27: (2) 95–117, 167–202, 281–308, 361–91.

———. 1901. The Physiographic Ecology of Chicago and Vicinity. *Botanical Gazette* 31 (3): 145–82.

Davis, M. B. 1980. Quaternary History and the Stability of Forest Communities. In *Forest Succession*, edited by D. C. West, H. H. Shugart, and D. B. Botkin, 132–53. New York: Springer-Verlag.

Davis, W. M. 1899. The Geographical Cycle. *Geography Journal* 14:481–504.

Drury, W. H., and I. C. T. Nisbet. 1971. Inter-relations between Developmental Models in Geomorphology, Plant Ecology, and Animal Ecology. *General Systems* 16:57–68.

Fernald, M. L. 1925. Persistence of Plants in Unglaciated Areas of Boreal America. *Mem. Am. Acad. Arts Sci.* 15:239–342.

Gleason, H. A. 1926. The Individualistic Concept of the Plant Association. *Bulletin of the Torrey Botanical Club* 53:7–26.

Goodlett, J. C. 1954. *Vegetation Adjacent to the Border of the Wisconsin Drift in Potter County, Pennsylvania*. Harvard Forest Bulletin No. 25.

Hack, J. T. 1965. *Geomorphology of the Shenandoah Valley, Virginia and West Virginia, and Origin of the Residual Ore Deposits*. U.S. Geol. Survey Prof. Paper No. 484. Washington, D.C.: USGPO.

————. 1965. *Postglacial Drainage Evolution and Stream Geometry in the Ontonagon Area, Michigan.* U.S. Geol. Survey Prof. Paper No. 504-B. Washington, D.C.: US-GPO.

Hack, J. T., and J. C. Goodlett. 1960. *Geomorphology and Forest Ecology of a Mountain Region on the Central Appalachians.* U.S. Geol. Survey Prof. Paper 347. Washington, D.C.: USGPO.

Odum, E. P. 1969. The Strategy of Ecosystem Development. *Science* 164:262–70.

————. 1983. *Basic Ecology.* Philadelphia: Saunders College Publishing.

Whittaker, R. H. 1975. *Communities and Ecosystems.* 2nd ed. New York: Macmillan.

7 Secondary Succession

In 1942 I took a course in ecology that used the standard text by Weaver and Clements on plant ecology (1938), which described the theory of succession as consisting of two important components: primary and secondary succession. I have alluded to portions of both of these components in previous chapters, but I feel it is important to address the theory as a whole and point out where it conflicts with Darwinian natural selection.

The scientific concept of succession is an abstraction of a higher order than simple recognition of the facts that: patches of vegetation vary from place to place; time sequences among patches can be found on river floodplains, sand dunes, or old fields; and the occupants of the patches can be arranged by categories of height, number of species present, and the rate at which plant material is produced.

Almost everybody has his or her own special meaning for "succession," akin to what Humpty Dumpty recognized in his discussion of definitions in Lewis Carroll's *Through the Looking-Glass:*

> "There's glory for you!"
>
> "I don't know what you mean by 'glory,'" Alice said.
>
> Humpty Dumpty smiled contemptuously. "Of course you don't—till I tell you. I meant 'there's a nice knock-down argument for you!'"
>
> "But 'glory' doesn't mean 'a nice knock-down argument,'" Alice objected.
>
> "When *I* use a word," Humpty Dumpty said in rather a scornful tone, "it means just what I choose it to mean—neither more nor less."
>
> "The question is," said Alice, "whether you *can* make words mean so many different things."
>
> "The question is," said Humpty Dumpty, "which is to be master—that's all."
>
> Alice was too much puzzled to say anything.

The very ill-defined nature of succession allows it to enrich nearly everyone's perception of nature. To many, it is simply change in vegetation and is used only to describe: the sequence of flowers from spring through summer into fall, the zones of vegetation evident on mountainsides, or the zones of vegetation from the tropics to the Arctic. Foresters and wildlife biologists in their practical applications, use "successional habitats" or "second growth" for almost any vegetation on the spectrum, from grassy fields, brambles, and brush to thickets and stands of saplings that "invade" cutover or burned-over land, blowdowns, old fields, river valley thickets, and sand dunes.

But succession has additional implications: the succession of fossil forms in geological ages, the "stages of evolutionary progress," the succession at the end of a dynasty inevitably involving a struggle for power. Succession comes to imply replacement and successful competition. For example, it has been applied to the zones of aquatics through emergent sedges to bushes and forest on the margin of a pond; the zones of plants on sand dunes, from the outer beach to the back dunes; the zones of plants from grasses and sedges, through willows and alders to trees on a river floodplain; the sequence of plants following a fire; the sequence of plants on abandoned farmland.

The Linnaean concept of morphologically defined species has had an important impact on how botanists think about vegetation. Plant ecologists accepted plant communities to be vegetation's equivalent of species —entities whose morphological features clearly distinguish them from neighboring forms. In this context Cowles emerges as the giant who founded the American school of ecology. He introduced the idea that vegetation is dynamic, in contrast to the idea then current in Europe that vegetation is static. Cowles studied and described the vegetation of the sand dunes on the south shore of Lake Michigan (1899, 1901). He identified the plant community as the unit of study and determined what he considered to be a sequence of vegetation, from the sparse grasses along the lakeshore, through brushy hollows, to forests of pine and oak or beech and maple on the dunes farthest inland.

Clements identified the long-term development of vegetation that occurs under natural conditions on geologically young surfaces such as mountains, river floodplains, and sand dunes as "primary succession" (1916). Clements hypothesized that during primary succession pioneer species colonize bare ground and, as they die, add organic material to the soil. As the soils deepen and increase in organic content, additional species colonize until the vegetation reaches its maximum development, the climax, which remains at equilibrium on the reduced relief of the oldest topography.

An important component of this theory is that species at each stage of the succession *prepare the way* for the species that will replace them. Early successional species alter their environment to make it better for subsequent species and worse for themselves. While individuals and species are replaced, the community as a whole moves toward some stable equilibrium ultimately determined by climate.

Clements applied the term "secondary succession" to changes resulting from windstorms, fire, and human effects because he considered them to be temporary deviations from the original climax. Since much of the initial "preparatory" work had already been done during the original or "primary" succession, secondary succession would take less time to achieve climax conditions. The actual amount of time required would of course depend upon how many "seral stages" were needed to restore conditions to equilibrium. It is important to note that inherent in this world-view is the notion that balance is the fundamental nature of things, and anything that disturbs this balance is considered "unusual" or an aberration that will be eliminated by the self-correcting mechanisms of the community.

The degree to which this view came to dominate much of ecology is evident in a paper by MacArthur and Connell published fifty years after Clements's formulation (MacArthur and Connell 1966): " . . . a clue to all of the true replacements of succession: *each species alters the environment in such a way that it can no longer grow so successfully as others.*" Twenty years later the notion of a climax equilibrium was alive and well in a widely used textbook in general ecology: "The climax is recognized as a steady-state community with its constituent populations in dynamic balance with environmental gradients" (Krebs 1985). Today, many environmentalists still hold onto its tenets, although many scientists have abandoned the essential elements of the theories.

[Editor's note: The notion of climax equilibrium is still going strong, as is evidenced in a recent ecology text by Rickleffs (1993): "Once forest vegetation establishes itself, patterns of light intensity and soil moisture do not change, except in the smallest details, with the introduction of new species of trees. . . . At this point, succession reaches a climax; the community has come into equilibrium with its physical environment."]

THE ORDERLINESS OF SECONDARY SUCCESSION

Secondary succession in abandoned fields has been widely used to illustrate succession in high school and college textbooks. Changes in vegetation in old fields have been observed on an almost infinite number of abandoned farms nearly everywhere in the eastern United States during the last 150 years.

Once a farm or garden is abandoned, the first plants that appear on the site are small, herbaceous, and fast-growing annual or biennial garden weeds. Perennial weeds follow the annuals, then woody brush overtops the wildflowers. Soon, trees appear—usually eastern red cedar and gray birch on sandy soils in southern New England, white pines in central New England and paper birch or white spruce in northeastern New England. Variations on the theme are common. Eventually, obvious change slows and "succession stops."

These changes have generally been explained by plant ecologists in the way that Whittaker did in describing succession on an old landslide: "One dominant species modified the soil and the microclimate in ways that made possible the entry of a second species, which became dominant and modified [its] environment in ways that suppressed the first and made possible the entry of a third dominant, which in turn altered its environment" (1975). But do the early species facilitate the establishment of the later arrivals, or, once established, do they just live out their lives until displaced by plants that overtop them? Where do the species come from that appear in succession? Is a particular stage in succession the adaptive habitat of particular species or do these species occupy other, geologically defined habitats, from which they expand their ranges opportunely and temporarily?

DIRECT EVIDENCE FROM ABANDONED FIELDS

McCormick examined the effects of early pioneers on succession in an old field, and his observations indicate that the early arrivals inhibited the establishment of later arrivals:

> Annual plants were removed as seedlings from some sections of a recently plowed field, but were allowed to grow elsewhere. According to reaction theory, an annual vegetation is necessary to "prepare the way" for perennial plants on such a site. By the end of each summer, however, perennial plants were several times as abundant on areas kept free of annuals. The biomass (dry weights) of individual perennial plants on the annual-free areas were many (15 to 82) times as great as those on areas with annuals. Many goldenrods, asters, black-eyed Susans, and other perennial plants flowered on annual-free sections, but were sterile on plots covered with annuals. . . . This experiment does not refute the general theory of the reaction mechanism. However, it does seriously question the reality of the theory and indicates that the theory was not valid for the early old field situation in which it was tested. (1968)

Note that McCormick, like most botanists at the time, was willing to consider his observations to be anomalies in a general law, rather than falsification of a preestablished hypothesis.

A long-term experiment in New Jersey by Pickett both explained how traditional old field succession may appear to occur and examined the detail of plant composition that refuted the theory. Pickett stated,

> Changes in species composition and cover were followed in an oldfield abandoned after plowing in the spring of 1960. Twenty years of data collected since then show the succession to be individualistic, that is, composed of broadly overlapping population curves through time. In general, the population curves exhibit long, persistent tails, indicating that, through this time span, succession is a process in which species that are present for much of the time become dominant at different times. Invasion and extinction are not the major mechanisms of community change. Bi- or multi-modal peaks were discovered in some species, . . . Many species which are important later in the sequence invade early. . . .
>
> . . . The community pattern is clearly a temporal continuum as has been long expected. . . . Examining only the *peaks* of the 10 or 12 leading species, i.e. ignoring the lower portions and tails of the species distributions, would allow the description of groups that correspond to the aspect dominant stages of succession often referred to. Clearly, however, examining the whole suite of species and their complete responses shows that the mechanism of community change is not one of sudden shifts between discrete compositional communities. (1982)

A major argument has been that some species, like alders and willows, prepare the way for "later successional" species, and hence that the delay in development of boreal forests could be attributed to the relatively recent retreat of the glacial ice. This seemed to have been confirmed by Cooper's studies of the colonization of raw soils in Glacier Bay, Alaska, where glaciers have been rapidly retreating during the last one hundred years (1923). Many textbooks have used this sort of observation to illustrate the "principle" that pioneer mat plants, shrubs, willows, and alders prepare raw soils for colonization by trees of the mature forest. But, recall from chapter 5, Sigafoos and Hendricks found that the time between the point when the ice left glacial moraines at Mount Rainier, Washington, and when tree seedlings became established was indeed short (1969). All of these examples show that different species respond in different ways, and that the reasons for events differ from place to place.

VEGETATION CHANGE
AS OBSERVED IN THE FIELD

I spent the first summer of graduate work taking courses at the Harvard Forest on the Massachusetts uplands in Petersham, about seventy-five miles west of Boston. When Harvard acquired the forest at the end of the nineteenth century the university planned for the forest to support itself through continuous cutting of its maturing stands of white pines according to principles of sustained yield.

It emerged from the detailed and long-term studies carried on at the Harvard Forest that the rolling hills and upland soils around Petersham had been cleared first by veterans of the American Revolution. By the 1840s, 85 percent of the land was cleared and in farms. Then in midcentury the economy of hill farms on the sterile upland soils of New England collapsed, and they were rapidly abandoned (Raup 1966). White pine seedlings, whose seed source was the trees that stood along fence lines and in woodlots, took over the abandoned farmland. These trees reached harvestable age at the end of the nineteenth century.

Sustained yield cutting of white pines preoccupied the early research at the Harvard Forest. This research became politically relevant when the Yale foresters were able to regenerate white pines at their experimental forest on the extensive sand plains of southern New Hampshire, while the Harvard foresters were unable to do so on the heavier soils of the uplands of western Massachusetts. The first director and the staff at the Harvard Forest expended a plethora of experimental efforts in order to maintain a sustained yield of white pine lumber, only to meet with perennial frustration. It became increasingly clear that seedling hardwoods had established root systems and grown up under the pines during the years of the original white pine stands. These seedlings were cut back, but after the mature white pines were cut the hardwoods sprouted from stumps. Because the hardwoods already had an extensive root system, they rapidly overtopped volunteer or planted white pine seedlings. The labor costs necessary to suppress the hardwood sprouts could not be justified economically.

The failure and frustration culminated in 1938, when a major hurricane crossed Long Island and passed up the Connecticut River valley, virtually eliminating the remaining mature growth of white pine. At that point, Hugh Raup was appointed director of the Harvard Forest. The 1938 hurricane not only ended attempts to reestablish white pine, it also gave Raup the opportunity to establish his own program of research. His central philosophy was to ask "naive questions." In this case he asked what the forest had been before it had been cleared by the settlers.

The summer that I was at the forest, Raup was planning a new program of research with the help of Steve Spurr, a professional forester. Spurr identified the oak-hickory stands on dry, coarse soils on south-facing slopes as successional, and he reserved the category *mature* for beech, yellow birch, paper birch, and sugar maple stands, all of which require moist, fine-grained soils (1950, 1956). These trees are indeed the most shade tolerant species. Over most of the slopes, Spurr pointed out, "two species are practically omnipresent. Red oak and red maple are equally prominent on all successional stages and one or the other is prominent on all sites. . . . Both exhibit a marked relationship to soil moisture, red oak being most frequent on the drier, and red maple on the wetter, sites" (1956).

Actually, most of the second growth covering the uplands consisted of patterns and mixtures of sugar and red maples, red oaks, white ash, paper birches, white pines, hemlocks, black birch, and some hickories. The patches seemed to vary strikingly from place to place, here approaching the transition hardwoods, there the northern hardwoods.

Spurr was already thinking in terms of delineating species distributions by site preference, using five categories of soil moisture. This eliminated the need to use the categories of successional stands that most foresters use: pioneer, transitional, and late successional. For example:

> Of the other species, the various birches are the most important. Paper birch is best adapted to the drier sites, black birch to the average well drained sites, and yellow birch to the wetter sites. White ash is locally important on the imperfectly drained soils, but shows little persistence into late successional stages. White oak is important in all successional stages

Paper birch, pen-and-ink botanical drawing, 1941.

on the drier soils, while hickories are somewhat less xerophytic, being found on the well drained as well as on the very well drained sites. Beech and sugar maple are occasionally found on the intermediate sites; while red and black spruce, tamarack, elm, and black gum are found on the wettest sites. (1956)

Note that this sort of description emphasizes the relationship of individual species to their environment, rather than stressing interrelationships between species.

Egler offered two testable alternative hypotheses focusing on "process" in successional theory (1954): relay floristics (each community alters the site in ways that allow others to colonize and displace it); and initial floristic composition (species colonize the area in a scramble, and some appear earlier because they disperse and grow more quickly). It is important to note the differences in process involved in these two hypotheses. In relay floristics early species "foul their own nest," creating an environment where they can no longer persist. Under initial floristic composition, seed dispersal and seed sources become the critical factors. In the 1950s, Egler offered a $10,000 prize (republished in 1975) to anyone who presents confirmable evidence that relay floristics occurred. Over the last forty years, no one has submitted a claim for the prize.

Under Darwin's theory of natural selection, individuals possessing heritable traits that enhance their survival and reproduction are likely to persist and spread. Individuals lacking these traits will die out. It would seem that the theory of relay floristics violates this process or at least consists of an easily circumscribed sequence. There would be strong selection pressures favoring members of "early successional" species to suppress their successors, and strong selection pressures opposing those individuals that enhanced the growth of competitors.

THE SECOND GROWTH: WEED SPECIES OR OLD GROWTH?

The vegetation that grows in a disturbed area is often patronizingly referred to as "second growth." Raup pursued the identity of such plants, and studies by his student Carlson showed that the species that follow the old-field "pioneers" deserve more respect. Raup and Carlson compared the distribution and abundance of the trees found during their study in the Harvard Forest with information recorded in one of those historical gems that occasionally come to the aid of a biologist interested in history (1941): Whitney's *History of the County of Worcester* written in 1793. Whitney had listed the trees found in remaining stands of upland forests and the tall trees that grew along the fence lines that had been left when the farmer cleared. Through this comparison Raup and Carlson found that the species composition of the second growth closely resembled that of the forests that covered Petersham when the first white settlers arrived.

Raup and Carlson compared the distribution and relative frequency of tree species in contemporary stands with that in the precolonial stands.

$10,000 Challenge

Challenge, to any believer in "plant succession to climax."

I, *Frank E. Egler,* hereby and herewith agree to wager any sum up to

TEN THOUSAND DOLLARS ($10,000)

against an equal amount, the money to be donated to a non-profit organization scientifically investigating the subject of Vegetation Change under natural or seminatural conditions, thru a period of more than 25 years, if any such believer will produce the evidence, either from the published scientific literature, or from unpublished research.

I stipulate that such research must support the Belief that natural and seminatural Vegetation change is a cause-and-effect phenomenon of ingoing and outgoing populations of plants, involving at least five stages, as indicated in diagrams published by me, in the sequence referred to as classical "Relay Floristics." Any contender will give advance notice in writing. He will prepare to submit all evidence in writing within six months of that time to a Commitee of Six Judges composed of ecologists Roland C. Clement, William H. Drury, William A. Niering, Ian C. T. Nisbet, and any two others they may appoint. The decision by the Judges will be reached within six months of the date of submission of the evidence.

Figure 2. In the more than forty years between the first publication of this challenge and his death in 1997, Dr. Egler's challenge was not met. (Quoted from Appendix VII in Egler 1975.)

When they compared the patterns of trees that had volunteered as second growth with the patterns of trees on land occupied by the linear descendants of trees that stood before European settlement, they found the patterns to be the same. The "old growth" matched the "second growth," even in the case of trees that followed old-field pines on plots that had been cleared, stumped out, planted to gardens, turned over to pasture, and finally abandoned.

This study indicates that once the white pines, which were in fact opportunistic old-field weeds, had lived out their lives, the tree species best adapted to the site grew back on their native habitat. The "best adapted species" are those of the wilderness forests, the same ones found by the first European settlers coming into the country. Obviously, this understanding markedly reduces the "turnaround time" involved in the story that regional forest development takes many thousands of years to reach climax.

RARE EVENTS OF EXCEPTIONAL INTENSITY

Over the years, Raup became more and more convinced that natural disturbances are so frequent that there is not sufficient time available for the sequence of generations implied in the theory of succession. He wrote:

> Wherever old American forests dating back to pre-settlement time have been studied historically they have failed to satisfy the requirements of the self-perpetuating "climax." One of the most important of these requirements is that the trees shall be all-aged. However, a universal feature of our old forests is that they are even-aged or have one or more well-defined age-classes in them. This phenomenon is known in so many parts of the continent, and in so many types of forest, that we cannot ignore it. We know of no way to account for it other than by the occurrence in pre-settlement time of disturbances that destroyed or decimated whole forest stands. (1964)

Henry and Swan found supporting evidence for Raup's view in the old growth forest of the Pisgah Tract in Southern New Hampshire (1974). They determined the age of stumps, sections of fallen logs, and even buried wood. They showed that the major species in this patch of old growth seeded during two exceptionally intense disturbances, a disastrous fire in 1665 and a major windstorm in 1938. As they put it: "We can say that tranquillity does not appear to be an important mediator of change, but that external events (fire and wind storms) are extremely important. The vegetational composition on one site may change considerably over time, and studies that examine compositional change associated with disturbance may provide a key to predicting its progress."

Periodic windthrow may serve a role similar to that performed by fires and agricultural clearing. Windthrow increases sunlight in small glades where several trees have fallen. Herbaceous plants may seed into these patches, as may tree seedlings.

Stephens's 1956 study in the Harvard Forest showed how the effects of windthrow have permeated the history of the forest. Stephens estimated the ages of trees in a patch. His counts showed first that the diameter of stem was not simply related to age. Second, and equally important, he showed that tree ages came in clusters, and each cluster was associated with some identifiable "disturbance," such as fire or a windstorm. He found evidence of powerful disturbance sometime between 1400 and 1500 and in 1635, 1730, 1750, 1815, 1851, and 1938. Note the long period without serious storms during which many inland New Englanders came to think that big storms were abnormal.

The mounds from trees felled by the oldest hurricanes were barely detectable on the surface, but Stephens found eighteen of them. He made a map of the locations and age classes of the fallen trees of the past, using it to draw in the area subjected to windthrow. The average areas and age classes turned out to be very nearly the same for each mound, however long ago the storm had come. The trees uprooted in the earliest recorded storm weren't much bigger than those later victims. The disturbances occurred well within the life expectancy of the major constituents of the "old growth stands." It seems likely, then, that few forest trees in the northeast live out their life expectancy, let alone provide for truly successional replacements.

Later, Oliver reworked Stephens's data and showed that the size of the area disturbed (the number of mounds created by each event) affected the species that became established in the patch (Oliver and Stephens 1977).

This suggests that the chance to win a place in the canopy may be a lottery in which failure may be the fate of most of the progeny of existing
species. Species are sorted out in response to local differences in soil, microtopography, subsoil structure, and exposure. The change is highly
probabilistic and indeterminate.

PATCH DYNAMICS

The ideas inherent in Oliver and Stephens's work on small patches of forest that are disturbed have been developed primarily by Pickett and
White into a more widely recognized phenomenon: patch dynamics
(1985). Following the theory of patch dynamics, death strikes individual
trees in the canopy capriciously, and when the individual dies, it falls and
opens up a glade. Then a scramble ensues until another individual that
has the "right stuff" and the luck to be at the right place at the right time
establishes itself.

Pickett and White's ideas represent a concept that has come to be identified as the fine-grained study of "disturbance" in "natural systems." I
have come to think of this as being a sort of calculus, in which change is
reduced to minimum space and time for analytical purposes, with the idea
that once the major themes have been identified, they can be applied to
vegetation in general.

When ecologists turned aggregations of plant species into "communities," they created the need for a theory to explain why so much of the vegetation is intermediate between the "pure types" that they conceived as
pigeonholes. This need is filled by the theory of succession. In a similar
fashion Pickett and White need a theory of succession on a micro scale.

As long as people regard death as disturbance and accept the idea that
a tendency in the direction of equilibrium is universal, we will need a
theory of disturbance and succession. But I don't think that either is valid
as an abstraction. For those who believe that natural communities are orderly, succession has been needed to explain why disorder is so pervasive.

When "disturbance" has been reduced to the area under an individual,
it becomes synonymous with death. Death has long been recognized as

necessary for life, but its pervasiveness can seem remote when the generation time of the individuals is several times our own. Few zoologists would take seriously assertions made about population phenomena in chickadees unless they were based on observations over several chickadee generations. Unfortunately, in much of ecology, the absence of evidence does not lead to skepticism regarding the underlying assumptions of theory. The degree of belief is shown clearly in Watt's comment: "On account of the lag in response or repeated and fortuitous checks to normal orderliness, or the long delay between the incidence of the disturbing factor and return to a possible normal, anything like phasic equilibrium may rarely be achieved even although the tendency is always in that direction" (1947).

The colonization of an old field demonstrates the tremendous potential for seed production and dispersal in the species involved. When an opportunity is offered, many become established; otherwise they die. The ordinary trees that push into gaps in the bushes or among the pine stumps are the species of the old-growth, wilderness forests. It is important to realize that romanticizing wilderness distracts attention from the impressive performance of ordinary species.

Henry and Swan found no suggestion of a trend toward an equilibrium hardwood stand–type in Harvard Forest, even over a century free of disturbance (1974). They found that the tree species that seeded into the large blowdowns held on until the next disaster. Hack's work that we saw in chapter 6 indicated that local details of the shapes and direction of slopes, relative soil moisture, and nature of the mineral soil, together with a generous dose of chance and historical accidents, helped to define species composition.

Egler's challenge to show that later successional types appear after the early ones have lived out their lives has not been met. It seems worthwhile to search for yourself in a New England field abandoned within a dozen years or so to see whether seedling birches, aspens, and pines (even red maples and oaks) are already showing up among the meadowsweet and sumac bushes.

The most important idea to carry away from this chapter is Egler's: all plant species that will be seen in an old field appear soon after the site has been abandoned. Egler's idea is a challenge to the fundamental assumption that certain species prepare the site so that others can more readily colonize. He says that succession is a series (to us) of individuals that live out their troubled lives, then die. The classical theory of succession sorely needs evidence of relay floristics—that certain sets of species actually replace each other. There is no evidence from the field to support Watt's idea that in the absence of disturbance, change occurs steadily, moving the vegetation towards an equilibrium. This remains an unfounded article of faith.

The classical concept of succession has become one of the most frequently cited examples of an ecological explanation for "the way the world works" and is featured in virtually every introductory biology textbook from grade school to college. Discussion of possible alternatives are not mentioned or are relegated to brief footnotes.

Successional theory has become a key component of much of the conservation movement. If natural areas—that is, those "uncontaminated by humans"—tend toward some sort of stable climax community through a predictable sequence, the best strategy that a conservationist can adopt is to remove human influences from a region, and wait for the inevitable reestablishment of the climax.

A first sign that an ecologist accepts the developmental model is a reference to "a relatively uniform habitat of regional extent." This suggests lack of interchange with neighboring, incrementally different species associations, hence the community is a closed system. Without exchange with neighboring populations, the local population must regulate its own numbers. Hence, stable animal populations must have inherent mechanisms that regulate their own populations. Again, because the habitat is assumed to be uniform and of considerable extent, subdominant individuals do not have places in which to escape or to hide in less preferred sites. Lacking heterogeneity of habitat for refuges, subdominant species develop mechanisms to avoid competition with their species neighbors.

Thus the acceptance of closed systems and developmental models in animal ecology is manifest in emphasis on density-dependent mechanisms of population regulation and on primacy of competition in limiting or determining the species composition of the community. We shall examine the consequences of these manifestations in subsequent chapters.

SOURCES

Carroll, L. 1971. *A Norton Critical Edition: Alice in Wonderland.* Edited by D. J. Gray. New York: W. W. Norton.

Clements, F. E. 1916. *Plant Succession: An Analysis of the Development of Vegetation.* Carnegie Inst. Washington Publ. No. 242:1–512.

Connell, J. H., and R. O. Slayter. 1977. Mechanisms of Succession in Natural Communities and Their Role in Community Stability and Organization. *American Naturalist* 111:1119–44.

Cooper, W. S. 1923. The Recent Ecological History of Glacier Bay, Alaska. Parts I, II, and III. *Ecology* 4 (2): 93–128; (3): 223–46; (4): 355–65.

Cowles, H. C. 1899. The Ecological Relations of the Vegetation on the Sand Dunes of Lake Michigan. *Botanical Gazette* 27:95–117, 167–202, 281–308, 361–91.

———. 1901. The Physiographic Ecology of Chicago and Vicinity. *Botanical Gazette* 31 (3): 145–82.

Drury, W. H., and I. C. T. Nisbet. 1971. Inter-relations between Developmental Models in Geomorphology, Plant Ecology, and Animal Ecology. *General Systems* 16:57–68.

———. 1973. Succession. *Journal of the Arnold Arboretum* 54 (3): 331–68.

Egler, F. E. 1954. Vegetation Science Concepts. 1. Initial Floristic Composition. A Factor in Old-Field Vegetation Development. *Vegetatio* 4:412–17.

———. 1975. *The Plight of the Rightofway Domain.* Mt. Kisco, N.Y.: Futura Media Services.

Henry, J. D., and J. M. A. Swan. 1974. Reconstructing Forest History from Live and Dead Plant Material—an Approach to the Study of Forest Succession in Southwest New Hampshire. *Ecology* 55 (4): 772–83.

Krebs, C. J. 1985. *Ecology.* New York: Harper and Row.

MacArthur, R. H., and J. H. Connell. 1966. *The Biology of Populations.* New York: John Wiley and Sons.

McCormick, J. 1968. Succession. *VIA* 1: 22–35, 131–32. Student Publication of Graduate School of Fine Arts, University of Pennsylvania.

Oliver, C. D., and E. P. Stephens. 1977. Reconstruction of a Mixed-Species Forest in Central New England. *Ecology* 58 (3): 562–72.

Pickett, S. T. A. 1982. Population Patterns through Twenty Years of Oldfield Succession. *Vegetatio* 49:45–59.

Pickett, S. T. A., and P. S. White, eds. 1985. *The Ecology of Natural Disturbance and Patch Dynamics.* Boston: Academic Press.

Raup, H. M. 1964. Some Problems in Ecological Theory and Their Relation to Conservation. *Journal of Ecology* 52 (suppl.): 19–28.

———. 1966. The View from John Sanderson's Farm: A Perspective for the Use of Land. *Forest History* 10 (1): 2–11.

Raup, H. M., and R. E. Carlson. 1941. *The History of Land Use in the Harvard Forest.* Harvard Forest Bulletin No. 20.

Rickleffs, R. E. 1993. *The Economy of Nature.* 3rd ed. New York: W. H. Freeman.

Sigafoos, R. S., and E. L. Hendricks. 1969. *The Time Interval between Stabilization of Alpine Glacial Deposits and Establishment of Tree Seedlings.* U.S. Geological Survey Prof. Paper 650-B. Washington, D.C.: USGPO.

Spurr, S. H. 1950. Stand Composition in the Harvard Forest as Influenced by Site and Forest Management. Ph.D. diss., Yale University.

———. 1956. Forest Associations in the Harvard Forest. *Ecological Monographs* 26:245–62.

Stephens, E. 1956. The Uprooting of Trees: A Forest Process. *Soil Sci. Soc. Am. Proc.* 20:113–16.

Watt, A. S. 1947. Pattern and Process in the Plant Community. *Journal of Ecology* 35:1–22.

Weaver, J. E., and F. E. Clements. 1938. *Plant Ecology.* 2nd ed. New York: McGraw-Hill.

Whittaker, R. H. 1975. *Communities and Ecosystems.* 2nd ed. New York: Macmillan.

$\mathcal{8}$ Studies in Evolutionary Biology

I have described some geological processes by which landscapes are sculpted and unconsolidated materials collected and distributed, as well as the effects that some botanists have found these processes to have on the distribution and abundance of plants. I have also described the theory of succession by which many ecologists account for patterns of vegetation per se with little acknowledgment of the effects of geological processes.

For these ecologists the theory of succession has become the central organizing theory of community ecology. Yet I have also pointed out that some plant ecologists have criticized this theoretical basis and have found evidence in the distribution of plants that contradicts most of the central tenets of succession. My discussion of this has been carried out with limited reference to the theory of natural selection, Darwin's theory of the process that leads to evolutionary change. This theory has been identified by most biologists as the most important theory in biology. As Dobzhansky said, "Nothing in biology makes sense except in the light of evolution" (1973).

In this chapter I will examine the various principles contained in the idea of evolution by natural selection. I encourage the reader to contrast the conclusions that will be reached by strict adherence to these principles with those contained in the theory of succession. I will begin by explaining how I was taught evolution, and then present some examples of how the theory of natural selection has been applied to studies of ecology.

POPULATION GENETICS AND NEO-DARWINISM

When I was at Harvard, the university was dominated by laboratory zoologists who took a reductionist attitude toward the developing theory of evolution. These ideas also resonated with the equilibrium notions represented by the Lotka-Volterra population equations. Emphasis was placed on the segregation of genes and the regularities illustrated in the Hardy-Weinberg equation in the absence of selection pressure. The focus was on the usefulness of equilibrium equations and the then-rapidly growing field of biological statistics in interpreting experimental data. The clear subliminal message was that it must take considerable competitive ability in a new species' population to break into all those equilibrium states, invade existing communities, and overlap with the geographic ranges of existing species.

The bright light in this was a splendid class in vertebrate paleontology that I took from Alfred Romer. Romer was interested in the "vertical" effects of evolution, the manifestations of its effects in the fossil record that showed changes in anatomy. Romer emphasized the nondirectedness of evolution and the confusion introduced when ideas like evolutionary inertia are introduced.

Romer made two other important points that stuck. First, all parts of an animal and all stages of its life cycle are affected by evolutionary change; second, some parts of a body may change very rapidly over a short time, while others will stay unchanged—a form of mosaic evolution. These ideas led me to examine the role of natural selection in the every-

Ruffed grouse, pencil drawing from
sketchbook, 1937.

day lives of the plants and animals I saw around me. This examination has
continued throughout my career.

NATURAL SELECTION: A PASSIVE PROCESS

Darwinism is not a single theory. It consists of a "bundle of concepts." An
important number were not clear in Darwin's time but have become com-
prehensible since then. Indeed, Darwin misunderstood several important
ideas that were clarified by later work (for example, he gave a degree of
importance to the effects of use and misuse and the blending of the blood
between the lines of inheritance represented by parents). In the following
list I summarize the major concepts.

1. Descent with modification replaces the idea of Creation or Divine
    Plan, making the existence of similar characteristics in markedly
    different groups a simple matter of explanation.
2. Population thinking, in which a population is made up of individ-
    uals expressing a variety of characteristics, is important.

3. The production of great excess of young is assumed.

4. Natural selection and fitness are understood to require examination in an historical context. Selection does not anticipate demands of the habitat or meet challenges. Evolution does not guarantee improvement. Selection does not give direction to evolutionary change. Any particular individual is not changed by selection. Those individuals who are more successful in reproducing already have the helpful traits. Existing traits are the secondary consequence of recent historical events and contexts that made certain qualities helpful in the local context.

5. The germ plasm provides the inherited material and occurs in the form of discrete particles, each a unit by itself. While new genetic material appears by mistakes in copying during meiosis, most of such differences are irrelevant or neutral. The major source of novel variation on which selection acts is recombination of already existing variation.

6. There is strict separation of germ cells from somatic cells at the earliest stages of embryology, so that there is no influence on future germ cells of the somatic cells in a particular generation.

7. Selection acts on all stages of the life cycle and cannot act on one part of the body without having secondary effects on many others.

8. The genome is selected as a whole; genes are not selected individually. The holistic expression of the genetic parts is the individual's phenotype.

9. Chance events that affect the survival and breeding success of individuals provide additional avenues by which traits can persist that are neither immediately advantageous nor disadvantageous (neutral).

A central assumption of the theory of natural selection is that there are reasons for an individual organism's characteristics that are driven by inheritance and the environmental "experiences" or selective events experienced by its ancestors. Even those traits that vary in response to the individual's own experience must have some inherited basis that set the extremes to which they will vary.

Some individuals do better than others, and their traits will be found in their descendants. If they have more descendants than their neighbors, their traits will spread. Individuals, once established among those that have access to resources, produce a clear excess of young, many more than the habitat can accommodate. The number and identity of those that survive and succeed is a consequence of the effects of the habitat. In this way what we see should be recognized as a passive process—a consequence— rather than a purposive process as we imply when we use active progressive terms like "evolution" and "adaptation."

Unfortunately, many people have concluded that populations adapt to an environmental challenge by changing. For example, Dobzhansky observed that "the best way to envisage the situation is as follows: the environment presents challenges to living species, to which the latter may respond by adaptive genetic changes" (1973).

This sort of underlying assumption has led many ecologists to use terms such as "strategy" that imply purposiveness or goal orientation. I don't like the suggestion that the environment presents a challenge or that the *population* responds. The population, after all, is an abstract construct. The reality is individuals. To offer a homely illustration of "population change," think of socks as a moving population on peoples' legs. The "population" may appear to vary or to remain the same, but it is the membership that is constantly changing. If one's socks change from blue to red or gray, they have not been dyed; they are different socks. Changes in a population of plants or animals represent the different characteristics of different individuals. The change in the population is a secondary effect of replacement of individuals.

Selection causes the differential representation of traits from generation to generation. Variants exist; environmental forces and chance, as well as interactions among individuals of the same species, choose who will survive. So, as certain individuals are selected and others are denied, the relative abundance of traits in populations shift as habitats shift. It is important to realize that major influences come from outside the species, and that individuals do not change or "become adapted." Rather, the offspring of lucky individuals that have the proper traits under the circumstances are differentially well represented in future generations. The traits of the unlucky ones are differentially less-well represented as long as that selection pressure continues. It is not a case of right or wrong or good or bad, but of which traits do better in the here and now.

NATURAL SELECTION AND CLUTCH SIZE IN BIRDS

Much of the fieldwork done in the middle part of this century to examine the effectiveness of natural selection was performed by ornithologists. The reasons for this may lie in part in historical accidents, but the circumstance was also affected by the ubiquity of birds and the seeming obviousness of measurable factors subject to selection. Of central importance among these features was clutch size. The variation both within and between species in the number of eggs produced, and the relation of this variation to reproductive success, was, and continues to be, a major focus of research.

In 1947, David Lack started a long-term study of great tits in the English woodlands outside of Oxford to test whether reproductive rates reflect the action of natural selection. Lack's student, Perrins, continued the study and greatly expanded on the details. Additional studies of other species of birds provide comparative information.

A key question was what relationship existed between the number of eggs laid and the number of young fledged. Lack states that reproductive performance can be evaluated with some confidence in nidicolous birds—those species whose young hatch helpless and stay in the nest for many days (1954). Clutch size in most birds is far below the normal limit of egg production; if the eggs are taken when laid, the bird readily lays others in their place, but these other eggs are not laid unless the first are taken. The limit is not set by the number of eggs that the sitting bird can incubate, because hatching success—the proportion of eggs that hatch—is similar for clutches of all sizes found in nature, including those well above the median for a given species. The true answer, at least for nidicolous species, is that clutch size has been adapted through natural selection to correspond with the maximum number of offspring for which the parents can, on average, find enough food without seriously harming themselves.

In England, the common swift lays either 2 or 3 eggs, 2 more often than 3. An analysis of the mortality among nestling swifts in four colonies in southern England in 1946–50 showed that in broods starting with 2 young, 82 percent safely left the nest, whereas in broods starting with 3 young only 45 percent safely left the nest. Nearly all the losses resulted from starvation, broods of 3 suffering more than broods of 2 because the food was at first shared among 3 mouths instead of 2. As a result, broods of 2 gave rise on the average to 1.6, and broods of 3 to 1.4 survivors per brood (Lack and Lack 1951). Hence 2 was a slightly more efficient clutch size than 3; and had any swift laid 4 eggs, one may suppose that a yet higher proportion of the young would have starved, so that fewer, not more, young would have been raised per brood. While broods of 2 were more successful than broods of 3 in wet summers like that of 1948, in fine summers such as 1949 the average number of young raised per brood was higher from broods of 3 than of 2. This makes it understandable how individuals laying clutches of 2, as well as those laying clutches of 3, persist in the population.

Note that Lack's explanation of the existence of clutches of both 2 and 3 is different from an equilibrium model of gene expression such as the Hardy-Weinberg equation. Lack argues that one level of clutch size may be advantageous one year and the other advantageous another year under different conditions. Individuals expressing both forms exist because of the variability of nature. There is not an equilibrium condition at a single "best" answer.

Geographic Modifications of Timing and Clutch Size in Songbirds

Individuals within the same species may lay several clutches several times a year in some regions and lay one large clutch in others, as is evident among horned larks (Drury 1961). At the northern limit of their range in the eastern Arctic Islands, horned larks lay a single clutch of 6 eggs. Food is abundant for a few weeks after they arrive in June, and the time is short before the snows return in late August and cover insect larvae that the birds seek among the mat plants. The larks have to hurry through their breeding cycle.

In other parts of horned lark breeding range, such as the deserts of the Great Basin in the western United States, females lay only 2 eggs in a clutch. The food available in those places is sparse all the time. Rains are irregular and unpredictable. It might be a prudent practice to lay three clutches of 2 eggs (spreading out the risk), or it could be that the females cannot find enough food in any particular short time to make more than 2 eggs.

This same phenomenon is manifest in European robins. Lack showed that robins nesting in England, a large island with relatively moderate and damp climate, lay on average 5.1 eggs; those nesting in southern Finland and Estonia, an area of continental climates with strong contrasts between winter and summer food supplies, average 6.3. Birds from Spain lay on average 4.9, and those on the Canary Islands, small islands with little difference between winter and summer weather, lay 3.5 eggs. Of course, within each population some individuals lay more or fewer than average.

In Britain's more uniform climate, where winters are damp and raw and the summers only slightly less so, robins winter on the same holdings where they nest. With little contrast in the amount of food available in "unfavorable" winter and "favorable" spring, females are hard put to sequester enough "excess" food to make eggs. In contrast, in eastern Europe where the winters are dry and cold, most robins leave their holdings in winter. And when spring comes again, there is a bloom of insect life. This suggests that "risk capital"—food available—governs the amount of energy that can be expended on young.

Females cannot simply "look forward" to what they hope will be the conditions when the chicks hatch and fledge. They have to deal with the food at hand while they are laying. I found this to be clear in the timing of laying among killdeer, a plover nesting on farmland over much of the United States (Drury, unpublished data). Killdeer return to eastern Massachusetts in middle or late March, establish territories, and form pairs while the frost is still close to the ground surface. Females do not lay until late April or early May, soon after spiders and earthworms become active in surface soil and the birds can find earthworms in the damp, bare ground between the sodden windrows of last summer's weeds. Once the food supply increases, the females are free to reproduce.

Clutch Size in Seabirds According to Distance to Feeding Grounds

After his work in the late 1940s that established the concept of selection pressure to lay as large a clutch as possible, Lack became interested in why clutch sizes might be reduced. He identified patterns that illustrate the trend to lay fewer eggs among shore- and seabirds (Lack 1967). The guiding influence is the distance the birds must commute to find food and therefore how often they can carry food home. Herons and cormorants, which feed at relatively short distances (a few miles) from their nests, raise more young than do gulls and terns, which feed farther from home (perhaps ten to twenty miles). Herons and cormorants usually lay many eggs, between 4 and 8; gulls and terns usually lay either 2 or 3 eggs. Birds of the deep sea, like gannets, auks, petrels, shearwaters, and albatrosses, cross tens or hundreds of miles of ocean to their feeding grounds, and they lay a clutch of 1. The ways in which the members of a species distribute their energies differ markedly according to their ways of life.

Herons and cormorants have young that stay in or near the nest for several weeks, and they have an abundant food supply close to their nest sites. Females can collect enough food to lay many eggs in a short time, and they lay large clutches of, at times, more than 5 eggs. The parents can also bring many loads of food a day to their nestlings. The young are fed for a month to six weeks until they fledge, at which point they are on their own.

Herons have a cruel system for preventing the overcommitment that can result from laying more eggs than can be fed as nestlings. In an estuary, it is usually impossible to anticipate in April whether the bait fish will be "up" in July or will not come back at all. The parents start to incubate as soon as the first egg is laid. Consequently, the chicks from later-laid eggs are at a disadvantage in the contest for food, being younger and smaller than their older siblings. Their older brothers or sisters elbow them away when food is short, or they may be eaten or trampled into the bottom of the nest. In this rather savage, Procrustean way, the supply of food adjusts the number of fledglings raised.

Gulls and terns also have young that must be fed for several weeks. Among the terns, clutch varies with the distance from nesting island to

Plate 4. A common tern with chick. Photo by the author.

feeding grounds. Common terns, which feed relatively close inshore, near their nests, usually lay 3 eggs, but the third egg is characteristically smaller and the chick from it seldom makes it to fledging. It is no easy matter for common tern parents to bring enough food to 2 chicks. It seems as though the parents play a dirty trick on the third chick, using it as insurance against the failure of one of the first eggs.

Arctic terns and roseate terns feed at greater distances from their nests than do common terns, often traveling ten or fifteen miles away over shoals. Arctic terns tend to lay 2 eggs, with only about 12 to 15 percent laying 3, while roseate terns consistently lay 2. Sooty terns and bridled terns, tropical terns that fish far offshore in the beautifully blue but unproductive tropical seas, lay only 1 egg. The tropical terns must fly for hours, well out of sight of land, in their search for fish. These parents need to work hard to find enough food to feed even a single hungry fledgling.

REPRODUCTIVE STRATEGIES AND TELEOLOGY

Pianka suggested, teleologically, that reproductive adaptations may allow rapid recovery of numbers or increase competitive ability (1970). This ar-

Double-crested cormorant, pen and ink, circa 1965.

gument was evident in MacArthur and Wilson's deduction from their equilibrium theory of island biogeography (1967). They dedicate one chapter to reproductive strategies, segregating pioneer species—favored in the early years of island colonization—from equilibrium species. They call the "opportunistic" pioneer species "r-selected" species, that is, having high reproductive capacity, quick attainment of sexual maturity, broad ecological tolerances, and well-developed dispersal mechanisms. These adaptations allow a species to invade recently disturbed areas. According to MacArthur and Wilson, the early successional r-selected species are replaced by K-selected or "equilibrium species." K-species have less-well-developed dispersal mechanisms and are less well adapted to the rigorous conditions of open tracts, but are better competitors.

The equilibrium theory of island biogeography concludes that K-selected species will be able to survive for long periods on islands because their conservative strategies do not lead to rapid buildup in numbers that would result in damaging population crashes. K-selected species are stayers.

Gadgil and Solbrig examined these "strategies" and, without identifying the teleological connotations, still concluded: "The concept of r- and K-strategies is meaningful only on a comparative basis, there being no absolute criterion to determine whether an organism should be classed as an r- or K-strategist. The most important characteristic of a r-strategist is that it devotes a greater proportion of available resources to reproduction than a related K-strategist" (1972).

Whorled aster grows on the shaded floor of the spruce forests on islands along the coast of Maine. Where the stands of spruce are close and tall, individual stems of whorled aster are about a meter apart and only occasional stems produce three or four flowers. Where glades have been formed by felled trees, and flickering sunlight penetrates to the floor, the lucky stems of whorled aster put out six to a dozen flowers. Where the ground is drenched by sunlight from a large blowdown, the plants grow shoulder to shoulder and put out twenty-five or thirty-five heads. How should we interpret this "behavior"—is the whorled aster an r-strategist or a K-strategist?

The plants in the open are vulnerable to being overshadowed by raspberries, which seed into the sunlit openings and in several years overtop the asters. Some people might interpret the massive flowering as an adaptation to set seed for dispersal before life on the current site becomes intolerable. Alternatively, it could be that after the energy requirements needed for maintenance are met, plants can turn their surplus energy into flowers and seeds. If the plants in deep shade put out a lot of flowers, they would not be able to provide enough nutrients to the seeds. The plants that get more sun produce more flowering heads.

PROXIMATE AND ULTIMATE MECHANISMS

Ecologists are continually moving between attempts to establish the immediate or "proximate" cause of a behavior (the how) and the ultimate evolutionary explanation (the why). Nest site selection by Arctic terns illustrates the difference between proximate and ultimate. Where Arctic terns nest on gravel bars of Arctic rivers, they regularly place their nests next to a dense clump of wildflowers or willows. The stimulus of a conspicuous object that brings the response of making a nest-scrape is proximate. That is a how. The ultimate cause is that birds that nest next to clumps of perennial plants do so on patches of gravel that have not been flooded for several years and probably will not be washed out during the three weeks of incubation. In contrast, gravels lacking plants are much more likely to be in a stream channel and to be flooded in the next rainstorm. The pairs that nest next to a hummock of flowers will probably produce more descendants than those that don't. That is a why.

Proximate causes are psychological and physiological mechanisms. As an example, Hohn explained the aggressive behavior of breeding female phalaropes as being caused by the presence of high levels of testosterone, the hormone that affects aggressiveness in males (1967). Evolutionary biologists would call this mechanism a how explanation. The why is that aggressive behavior of females is related to their establishing territories and courting males. Female phalaropes lay several clutches and males incubate them. The exchange of roles is an adaptation that allows a female to lay many more eggs than she would be able to if she were paired to a single male in the short time during which food is abundant in the tundra pools of the far north. Midge and mosquito larvae occur in abundance when the birds come back, and this means that females can lay several clutches. They may have to, because Arctic foxes are constantly patrolling for clutches to steal. By incubating the eggs, a male protects his investment in the eggs he fathered. This allows the female to feed, regain her energy, and replace eggs taken by a marauding fox—that is the survival value.

Mechanistic explanations—the physiological basis—represent proximate causes. These are interesting studies, but they are not more important because they are "more fundamental" than explanations that involve the ultimate advantages individuals gain by having certain traits or doing certain things. But most physiologists and geneticists disagree, and their reductionistic philosophy is widely accepted.

UNITS OF SELECTION

Let me emphasize that the real things we can see are individual organisms, their anatomy, and cells. When we come to consider populations or species, we have embarked on the realm of progressive abstraction. We may refer to a population, but is it a reality in nature or the net effect of a number of secondary consequences of the activities of individuals? Similarly, a gene is a "place" on a chromosome that we can recognize by secondary effects of the activities that seem to be mediated by something at this place. Selection does not act on a population. Rather, changes in the characteristics of the individuals that make up a population are the secondary consequences of changes in what traits individuals carry.

Dawkins and many others have argued that selection acts on the genes, as the genes are the immortal material that is carried on from generation to generation (1989). Intact individuals, he argues, are merely the vehicles that carry the active agents in their temporary manifestation. Others have argued, and I agree, that genes cannot exist separate from the individuals they are in. The success of any gene depends not solely on itself but on all the other genes with which it joins together to make up an individual.

My thinking is influenced by how complex systems are and the aphorism that a chain is only as strong as its weakest link. I am reminded of the design of an antiaircraft weapons system. Such a system depends on a propulsion system for the missile, its guidance system, its system to detonate the ordinance, its radar guidance system, and computers. These must all work together and they are "selected" as teams. The military selection among several alternative systems does not include "trying" all the units in all possible combinations with each other. A system is likely to fail as a unit without its several elements being independently tested. Similarly, sets of genes that do work are tested as units: the individuals that carry them. The survival of individuals produces success in all of their components.

Other researchers, like many social scientists, are attracted to the idea that a group is selected as the unit. Wynne-Edwards argued that separate colonies of gulls would be selected against if they overcropped their resources (1958, 1959, 1962). Therefore the group developed ritualized ceremonies—mass displays—that either stimulate the group as a whole to breed or discourage marginal individuals if the population is pressing on its resources. This is an old argument that suggests the goal orientation of "strategies" or "for the good of the species." If individual gulls that "cheat" on the systems of restraint move into a well-regulated gullery, however, they will upset the order. One of the great values of Wynne-

Edwards's hypothesis was that it so infuriated a number of field biologists that they felt constrained to present contrary arguments that considerably clarified the consequences of selection.

Group selection would require a number of closed systems that operate like independent organisms on which selective forces could act. Genes, cells, or populations do not function as free-standing units. Selection acts on individuals, each containing a large body of information in genes that is well protected from meddling by mutation and recombination.

SOURCES

Bazazz, F. A. 1979. The Physiological Ecology of Plant Succession. *Ann. Rev. Ecology and Syst.* 10:351–71.

Dawkins, R. 1989. *The Selfish Gene.* New ed. New York: Oxford University Press.

Dobzhansky, T. 1973. Nothing in Biology Makes Sense Except in the Light of Evolution. *American Biology Teacher* 35 (3): 125–29.

Drury, W. H. 1961. Studies of the Breeding Biology of Horned Lark, Water Pipet, Lapland Longspur, and Snow Bunting on Bylot Island, Northwest Territories, Canada. *Bird Banding* 32:1–46.

———. Unpublished data collected in the 1960s in Massachusetts. Data on file in the author's library.

Gadgil, M., and O. T. Solbrig. 1972. The Concept of R- and K-Selection: Evidence from Wild Flowers and Some Theoretical Considerations. *American Naturalist* 106:14–31.

Grime, J. P. 1977. Evidence for the Existence of Three Primary Strategies in Plants and Its Relevance to Ecological and Evolutionary Theory. *American Naturalist* 111:1169–94.

Harper, J. 1977. The Contributions of Terrestrial Plant Studies to the Development of the Theory of Ecology. In *The Changing Scenes in Natural Sciences 1776–1976,* 139–57. *Academy of Natural Sciences,* Special Publication 12.

Hohn, E. O. 1967. Observations on the Breeding Biology of Wilson's Phalarope (*Steganopus Tricolor*) in Central Alberta. *Auk* 84:220–44.

Lack, D. 1954. *The Natural Regulation of Animal Numbers.* Oxford: Clarendon Press.

———. 1967. Interrelationships in Breeding Adaptations as Shown by Marine Birds. In *Proceedings of the XIV International Ornithological Congress, Oxford, 24–30 July 1966,* edited by D. W. Snow, 3–31. Oxford: Blackwell Scientific Publications.

Lack, D., and E. Lack. 1951. The Breeding Biology of the Swift *Apus Apus*. Ibis 93 (4): 501–46.

MacArthur, R., and E. O. Wilson. 1967. *Theory of Island Biogeography*. Princeton: Princeton University Press.

Mayr, E. 1991. *One Long Argument*. London: Penguin Books.

Perrins, C. M. 1979. *British Tits*. Glasgow: William Collins Sons.

Pianka, E. R. 1970. On R- and K-Selection. *American Naturalist* 104:592–97.

Wynne-Edwards, V. C. 1958. The Overfishing Principle Applied to Natural Populations and Their Food-Resources: And a Theory of Natural Conservation. In *Proceedings of the XII International Ornithological Congress, Helsinki 5–12 VI 1958*, 790–94.

———. 1959. The Control of Population Density through Social Behavior: A Hypothesis. *Ibis* 101:436–41.

———. 1962. *Animal Dispersion in Relation to Social Behaviour*. Edinburgh: Oliver and Boyd.

_9_ Population Balance, Equilibrium, and Density Dependence

In this chapter, I will address the behavior of populations in order to clarify some ideas that have been significant to the development of general ecological theory, even when abundant contradictory evidence was available.

A central assumption of the traditional theory is that populations constitute a separate level of biological organization. Indeed, populations do have their own characteristics that apply neither to individuals nor to "communities." Populations have a size and geographic distribution, a generation time and an age structure, a brood size and age at first breeding, differences in mortality or reproductive rates by age and sex, a rate of population growth or decline, oscillations and regional differences in population density, and differences in morphology and behavior, including the range of habitats selected by the membership. A key element in my approach to populations is that these characteristics are secondary consequences emerging from the traits that historical events have selected among individuals.

I will show that high densities of individuals do have many observable effects on breeding performance and mortality, but that these effects, while pervasive, are not necessary and sufficient causes of population regulation. I will show how much more complicated and interesting are the multitudinous influences that apply in ordinary contexts. These influences reflect adaptive advantages provided to individuals that produce excess young who can then disperse over heterogeneous habitats. They also reflect some forces that work independent of density, and other forces that result from the ability of some individuals to dominate others. These together define which and how many individuals are vulnerable. These differences in state of mind involve a shift away from Newtonian mechanistic regularity to Ecclesiastes's general law that time and chance happen to all things.

CLASSIC POPULATION THEORY

Many of the assumptions upon which traditional competition theory and systems theory are based are irrelevant or wrong. Principal among these assumptions is the notion that populations are inherently stable or vary only between prescribed limits, and that there must therefore be mechanisms that maintain a balance between the members of the population and their habitat. This idea, that populations are in equilibrium or balance, was stimulated by the rediscovery of the logistic equation for population growth in the 1920s. From it, Nicholson deduced that populations must be regulated and that regulatory mechanisms must respond to density (1933).

If emigration and immigration are excluded or are required to remain equal, and the population changes in size, the change must be equal to the difference between births and deaths. The mathematical model, the logistic equation, describes the curve:

$$\frac{dN}{dt} = rN\left(\frac{(K-N)}{K}\right)$$

It makes intuitive sense that population growth and change is primarily determined by the number of individuals in the population (N) at the start. Changes in that number (dN) as time changes (dt) are the consequences of multiplying the initial N by the rate at which individuals recruit offspring (r).

The logistic equation modifies this estimate of population change by introducing the term $[(K–N)/K]$, in which K is the "carrying capacity" of the local habitat. Carrying capacity is defined as the population size that the environment can maintain; it is the hypothetical equilibrium point.

The rate of increase $[dN/dt]$ is affected by $[(K–N)/K]$. As N gets *larger* a smaller fraction of offspring is recruited; as N gets *smaller* a larger fraction of offspring is recruited. So the factor $[(K–N)/K]$ is a feedback mechanism that regulates the numbers in the population in a "density-dependent" manner. When the population goes up, either birth rates go down or death rates go up, so that the population remains at equilibrium near the carrying capacity. By logical symmetry, the equation implies that when a population is low the pressures inhibiting growth are released and numbers increase again toward the equilibrium.

The appearance of mathematicians lent respectability to ecology, which flowered as systems ecology. However, they made many field ecologists uneasy, as if "serious musicians" had suddenly decided that they would set things straight in folk music.

Problematic Assumptions of the Logistic Equation

A central element in the logic of density-dependent population regulation is the equilibrium carrying capacity. In the logistic equation this complex idea is represented by a single integrative mathematical term representing the sum of environmental resources necessary to maintain the population at equilibrium. When this level has been exceeded, further increase is inhibited. The concept of carrying capacity has a number of defects, primarily because it was taken to imply a *single* carrying capacity, which led many ecologists to think of "normal" densities and "usual" situations.

The idea of carrying capacity presupposes a stable environment. As I have stated earlier, during the early decades of the twentieth century most people's thinking about natural conditions was permeated with norms and balances. Such assumptions of a homogeneous and stable environment have led many ecologists to believe that "healthy" populations have feedback mechanisms that keep the population stable near the carrying capacity.

The concept of carrying capacity cannot accommodate the common observation that significant variations in population size and density occur consistently in different parts of a species' range. At a detailed level, the nesting populations of herring gulls on some islands are very dense while those on neighboring islands are sparse. One might conclude that these different densities reflect differing carrying capacities. On many islands, however, densities vary within the island population by factors of ten. If that much local variation is explained by local differences in carrying capacity, then the concept has little value in understanding populations except in the abstract.

The concept of population regulation indirectly implies community stability. The processes that "regulate" have been referred to as K factors, clearly implying the logistic equation. The formula has the attractive characteristic of integrating a lot of unknown forces. This complex of ideas becomes, then, the ecological equivalent of Adam Smith's "unseen hand" in industrial economics.

Once the logic of the curve of population growth is formally represented in the logistic equation, the hazards of logical symmetry and determinism become evident. The determinism was built into the logic because the biologists involved held to the older concepts of balance of nature. The more subtle hazard of the formulation is that it suggests that at low densities factors will come into action that stimulate population increase, just as at high densities factors come into effect that depress further increase.

Norms are also implied by averaging data in life tables from many individuals of markedly different competencies. While the use of averages allows the biologist to neaten up sloppy data, such treatments obscure individual variation. What is worse, these techniques tend to distract students from paying attention to individual foibles, and in fact they encourage biologists to think that the individual differences are "noise" and that inquiry into the behavior of individuals is irrelevant.

The use of averages in the logistic equation and life tables suggests that good samples of rigorous field data will have narrow standard deviations. This hints at the Microcosm/Macrocosm superstition that a set of samples can be taken randomly that will be representative of the whole. This further suggests that the whole is made up of subsets of homogeneous and representative parts. But instead of a symmetrical "normal curve" suitable for statistical tests, most populations show distributions of numbers or densities that vary in ways that produce polymodal plots.

Population "Regulation"

The widespread adoption of the logistic equation as a good first approximation of population growth had a profound effect on the development of much of ecology. Ensuing discussion based on mathematical arguments convinced most theorists that unregulated populations would be subject to what are called Markov processes. Markov processes are the consequences of stochasticity. According to this idea, if fluctuations are free to run their course without restraint, a random run of several declines in a row will inevitably lower the population to zero and extinction. Hence, mathematically, population fluctuations must be contained or regulated if the population is to persist. The laboratory cultures reported by Nicholson and others indicated an orderly increase toward the asymptote around which equilibrium was assumed to vary. Interestingly enough, in almost every case something went wrong with the laboratory cultures at the assumed equilibrium point, or the experiments were broken off. Despite this, Nicholson still concluded: "The evidence for the existence of a balance is not confined to logical deduction from the known facts about animal populations existing in a state of nature. Experiments dealing specifically with populations prove that the latter do reach a state of stationary balance under constant environmental conditions" (1933).

Not everyone agreed with Nicholson. Two leading opponents of the idea of density-dependent regulation of populations were Andrewartha and Birch, who criticized density-dependence as a logical necessity. They observed that "the usual generalizations about 'density-dependent factors,' when they refer to natural populations, have a peculiar logical status. They are not a general theory, because . . . they do not describe any substantial body of empirical facts. Nor are they usually put forward as a hypothesis to be tested by experiment and discarded if they prove inconsistent with empirical fact. On the contrary, they are usually asserted as if their truth were axiomatic" (Andrewartha and Birch 1954).

Andrewartha and Birch were studying a variety of insect populations that managed to persist in spite of wide swings in the environment. The two felt that for many organisms non-density-dependent factors such as weather might have a profound effect on population size long before density-dependent influences such as competition and predation could come into play. While their ecology text remains a classic in the eyes of many biologists, their ideas were largely drowned out by the supporters of equilibrium and regulation. Max Solomon, an experienced and creative student of insect populations, agreed with Nicholson:

> The answer given . . . [by Andrewartha and Birch] offers us only a coincidence: that the time limit to the periods of rapid increase allows just the right number to be produced to offset, over a number of years, the mortality during the dry season. It seems extremely unlikely that this would happen, without the occasional regulating assistance of a density dependent process. . . .
>
> . . . To explain why the great irregular increases and decreases from year to year add up over a period to approximately zero, we seem to have only two alternatives, density-dependence at some point in the population cycle, or pure chance. (Solomon 1957)

Lack, a supporter of population regulation, said: "As Nicholson (1933) showed, the idea [of density-dependent population regulation] needed only a logical, not a mathematical, basis for its acceptance, since if an animal population continues to fluctuate in numbers over a long period between restricted limits, it follows that it is controlled by factors which tend to produce an increase after a low density and a decrease after a high density; otherwise it will either increase indefinitely or become extinct" (1966). Notice that both Lack and Solomon refer to *a* population or *the* population, as if each population were an isolated entity, a closed system, functioning by itself.

Nicholson's paper was widely accepted as demonstrating that at high population densities individuals must become more vulnerable to disease, predation, starvation, and interindividual competition and birth rates must decline, while at low densities (by logical symmetry) birth rates will increase and death rates will be reduced. Indeed, laboratory studies seemed to show that the growth of populations of insect larvae in artificial media slowed as numbers approached what was assumed to be the equilibrium. Other laboratory studies revealed a physiological mechanism for failures of reproduction in mouse populations at high densities (Christian and Davis 1964).

Field evidence of cyclic fluctuations of lemmings on the wet tundra, of snowshoe hares in the uniform vegetation of coniferous forests, and of voles on the simplified vegetation of agricultural land was taken to confirm the idea that such "simple systems" lack mechanisms of adequate population regulation. This general agreement of logic and experimental science demonstrating apparently homeostatic mechanisms provided an almost insurmountable obstacle to ecologists who chose to understand population changes as largely independent events, defined by chance, occurring in many partially isolated subpopulations, and potentially canceled out by low levels of exchange of individuals.

ALTERNATIVE EXPLANATIONS

Movements and Extinction

A changeable environment, where local extinctions are chronic, selects for individuals with characteristics that scatter their descendants widely. The importance of movement between multiple population centers in maintaining a regional population was brought home to me when I became a laboratory assistant and had problems with maintaining a laboratory culture of a small invertebrate, the water flea *Daphnia*. My cultures rapidly deteriorated in spite of everything I tried.

I wrote to the source of the laboratory cultures I was using, explaining my problem and asking how they dealt with keeping *Daphnia*. Their answer was a classic of simple practicality. Unfortunately I did not keep the letter, but it ran something like this: "We have several dozen oak barrels behind our storage shed to collect rainwater, and there is always a healthy population of *Daphnia* in at least one of them." Thus, this species can be (and in nature doubtless is) "protected against extinction" by the simple expedient of dispersing between partially isolated nuclei. Total extinction is likely to occur only when all nuclei "crash to extinction" simultaneously. The probability of extinction decreases exponentially with the number of nuclei. Given dispersal between different nuclei, the issue of "population regulation" becomes moot.

It would be difficult to pick a species that shows less regulation of numbers within its nuclei than *Daphnia*. In herring gulls, the numbers in each nucleus are limited (except on the largest islands) by territorial behavior, which in present circumstances usually promotes emigration of some individuals. With the addition of spatial limitations within the nuclei, it seems clear that the "rain barrel" mechanism suffices to explain long-term persistence of herring gulls.

As a result of increasing bird-aircraft collisions, I, along with other members of the Massachusetts Audubon Society scientific staff, was contracted by the Federal Aviation Agency and the U.S. Fish and Wildlife Service to make an extensive study of the population biology of herring gulls. We were thrust into a situation in which we had to examine what would control gull numbers. Were effects of density-dependent population regulation evident in New England? We looked for effects of density on reproduction and mortality in herring gulls.

As Wynne-Edwards noted, three characteristics affect the reproductive rate of a population: the age when birds first breed, the proportion of nonbreeders, and the clutch size (1962). For these to regulate a population, all should change as a population grows, but in the population of herring gulls we studied they changed little between 1940 and 1970, despite a fourfold increase in the total population. Indeed, the proportion of nonbreeding birds was higher, the average clutch size lower, and the number of young raised per pair lower than average on the outer islands in Maine, where the breeding density is lower than average. At the same time, the proportion of nonbreeding birds was lower, the average clutch size higher, and numbers of young raised higher on the islands in Massachusetts Bay where the breeding density is highest (Kadlec and Drury 1968a; Drury and Nisbet 1972). Hence these factors do not act to regulate the population density.

Most people assume that the number of gulls nesting on an island remains about the same over the years, and that gulls in all colonies are about equally successful in their breeding attempts. We found the overall average success to be just less than one chick per nest per year, but, more important, that variation around this "norm" is large. See table 1. It becomes evident that the young produced on small islands do not replace adult mortality, while most of the young produced on large islands must emigrate (Kadlec and Drury 1968b; Drury and Nisbet 1972).

The differences in success show an instructive pattern. The smaller gulleries on outer islands have more empty nests than do the large ones near fishports. Gulls nesting on outer islands are more sensitive to bad weather and late springs.

Censuses of gulls taken every twenty years since 1900 show that the numbers of gulls nesting on outer islands have decreased steadily over the last fifty years (Drury 1973, 1974). Even while gulls were increasing and expanding their range rapidly to the west and southwest, they moved inshore off the outer islands.

Our studies of gulls' feeding movements showed that the gulls who cause the aircraft collisions are attracted to city dumps. In addition, our data suggest that many experienced adults moved away from islands when egg destruction by humans or the presence of foxes disturbed them. Also, the numbers of gulls nesting in some older, less productive gulleries on outer islands have decreased while numbers have increased on new, more productive gulleries near the mainland. Our file of banding recoveries shows that individual gulls have moved in both directions.

Adults from different gulleries segregate somewhat into different feeding areas, and some of this segregation persists even in winter. Young migrate long distances and mix freely in winter. Winter mortality rates of first-year birds vary more than do those of adults and appear to be higher in those regions where gull densities are lower. Young birds now disperse more widely and winter farther north than they did earlier in the period of increase.

There are marked differences in average nest spacing and in breeding

Table 1 Variation and average productivity (chicks per nest) of herring gulls in colonies of different sizes along the New England coast between 1963 and 1969

Size of colony	Average productivity	Range
<40 pairs	.7	.2–1.3
40–400 pairs	.95	.5–1.2
>400 pairs	1.15	.7–1.4

Note the large variation in the range. Data from Drury and Nisbet (1972).

success among gulleries, correlated with food supply, nesting substrate, and other factors, but we have found no evidence of a consistent density-dependent factor acting to regulate the population. Indeed, breeding success is higher in larger gulleries, while winter mortality is higher at lower densities. We could say that at individual gulleries and in groups of islands within a bay, numbers are "regulated by density-dependent emigration," but the limiting densities vary widely between gulleries and between regions.

Dispersive movements between gulleries appear adequate to explain the persistence of herring gulls despite local catastrophes. If food were restricted, the population as a whole would behave as though restricted by density-dependent mortality, even though regional mortality should remain uncorrelated with density. So, there is no need to postulate intrinsic regulating mechanisms to protect the species against "overpopulation" or extinction. Unfortunately, current theories of population regulation do not apply to a species in which breeding densities are influenced by dispersive movements and winter densities are determined by migratory movements.

For those who consider "successional habitats" to be different from "stable habitats," it is important to discuss the role of movements of bird populations of "climax" habitats, in particular in mature forests. If the habitat is at climax (by definition stable and spatially uniform), then immigration and emigration may be expected to balance each other and it should be possible to treat "a sample population" as a microcosm.

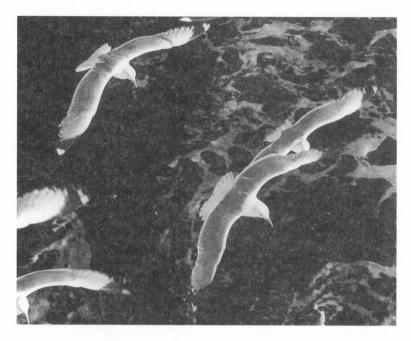

Plate 5. Herring gulls. Photo by the author.

The best-studied species is the great tit (Lack 1966). It is possible to argue, as Lack did, that this species is no longer found in its original habitat because the mature forests in the areas where it has been studied have been fragmented into "islands" in a matrix of less-favored habitats.

A counterargument should be clear to readers who have gotten this far: forest trees are not distributed uniformly within a "stand type," such that a randomly chosen sample can be considered to represent the whole. Islands of good, marginal, and submarginal habitat are distributed throughout virtually any "wild forest," and songbirds of even such a broad category as leaf gleaners recognize good habitat quickly. The Swedish ornithologist Svardson noticed that within mixed forests, spring migrant wood warblers moved silently through the stand until they came to one part where they stopped and sang vigorously (1949). Habitats that we judge to be homogeneous and uniform may appear to other eyes to consist of patches of very different quality.

Great tits are a hole-nesting species that depends in winter on seeds of characteristic climax trees (Perrins 1965, 1967). Breeding success and density vary substantially among different areas and across habitats (Kluyver 1951; Lack 1966; Dhondt 1971). In the isolated area on the island of Vlieland, in the Netherlands, adult mortality varied directly with the number of young produced (Kluyver 1966). Lack reported that annual fluctuations depended primarily on postfledging disappearance rates of young, and these were correlated not with density but with botanical factors. Kluyver (1951) suggested that some birds emigrated, and Dhondt and Huble (1968) confirmed this by capturing birds in neighboring areas. Kluyver (1966) found that 20 to 60 percent of the breeding population in one study area consisted of immigrants each year and recorded what were apparently immigrants even on Vlieland, isolated by twenty kilometers of tidal flats and shallow sea. In these respects, great tits, although noncolonial and nonmigratory, resemble the seabird species.

Different parts of the habitats used by great tits vary widely both in the density of the tits and in their reproductive success. In productive mixed woods, Kluyver found that great tits nested more densely and produced more fledglings per brood, occasionally producing second broods (1951). In pine stands or woods in which Scots pine made up a large portion of the canopy, the evenly spaced nest boxes were occupied less densely and the nests produced fewer young. When populations were low, few birds nested in pine woods. So local, marked variation in quality is evident in great tit habitats as in the habitats of the herring gulls.

Chronic Breeding Failure and the
Importance of Nonbreeding "Floaters"

The marine environment is highly variable in terms of food availability from one year to the next. In years of food abundance more herring gull pairs take part in breeding, and more pairs produce two or three chicks; in poor food years fewer pairs join the breeding population and fewer of them produce any young. See table 2.

Some seabirds suffer repeated breeding failures (Gaston and Nettleship 1981; Nettleship, Birkhead, and Gaston 1979). Studies conducted on kit-

Table 2 Percentage of herring gull nests on Block Island, Rhode Island, producing 0, 1, 2, or 3 chicks to fledging in a good and a bad year

	% nests producing chicks to fledging	
	Good year (1966)	*Bad year (1969)*
	N = 270	N = 215
0 chicks	11	40
1 chick	45	40
2 chicks	34	17
3 chicks	10	3

Data from Virginia and Merrill Slate (personal communication; see also Nisbet and Drury 1972).

tiwakes in Britain showed a geographic pattern to reproductive success (Heubeck 1988). The few productive nesting sites were separated by many miles of cliffs supporting only marginally productive ones. Drury, Ramsdell, and French found similar chronic failure between 1975 and 1978 in the kittiwakes at Bluff Cliffs near the Bering Strait (1980). Our studies were continued by Springer, Murphy, Roseneau, and Springer, who found that only during four years in the previous fifteen did production reach one chick per nest with eggs (1985). The birds failed completely in seven years. We found that 32 to 69 percent of the well-built nest platforms in any particular season did not ever hold eggs (Drury, Ramsdell, and French 1980). This suggests that in many years, females have difficulty gathering enough food to lay eggs. I think that these failing birds are an important part of population studies.

Human attempts to reduce gull population growth by destroying eggs can be used to measure the ability of local populations and groups of populations to persist in spite of breeding failure. This can be an important issue, as many environmentalists argue that every clutch is critical to the survival of a local population, and that disturbance will rid a place of gulls.

During Gross's 1951 egg-spraying program, which was intended to kill the embryonic gulls without breaking the eggs, crews sprayed about 350,000 gull eggs on Maine islands between 1940 and 1944. By 1945, the number of gulls in Maine was evidently decreasing, but the numbers in Massachusetts were increasing rapidly (Drury and Nisbet 1972). The numbers decreased on Maine islands too rapidly to be the result of mortality alone. It seems that spraying, like taking eggs and natural nest failure, stimulated gulls to move. The gulls did not sit idly by and suffer the consequence of the spraying program, which was, like that of taking eggs, chronic reproductive failure. They responded with their most effective weapon: moving. The historical evidence shows that the egg-spraying program accelerated the southward shift.

The program was subsequently expanded to the islands in Massachusetts Bay and south of Cape Cod, then to the islands in Block Island and Long Island Sounds. Between 1945 and the end of the program in 1953, half a million eggs were treated, yet the population of gulls nesting in southwestern Maine continued to grow, though slowly after 1945.

While Gross's program was inhibiting the breeding efforts of about 35,000 pairs of herring gulls in Maine, 35,000 pairs were breeding uninhibited in the Grand Manan Archipelago, just east of Maine. Young Canadian immigrants filled the ranks in the United States that had been thinned by the control program. The program failed because those who designed it did not recognize how readily gulls move and how elastic is the postfledging survival of newly fledged gulls.

In species that have developed a social structure there will inevitably be individuals who are excluded entirely from reproduction in the normal course of events. These individuals can either remain with their natal group in the hopes of eventually moving up the social hierarchy, or they can leave and become "floaters" roving from place to place until they die or can move into a new group.

According to Susan Smith, who followed black-capped chickadee floaters around campuses in Massachusetts, floaters live insecure lives (1985). They drift in and out of the flocks that circulate between a couple of bird feeders. But one of them usually moves in when a high-ranking bird dies in an established flock. Floaters have to play the odds of winning big in life's lottery while eking out lives of quiet desperation. When a hunter kills a deer, rabbit, or wolf, or when one of the chickadees at your feeder dies, a rover from that species happens by and takes over. As long as there are plenty of floaters waiting to step up, hunting has "no effect." But when the mobile reserves have been cleaned out, the population crashes.

These arguments led me to think that theories of density-dependent population regulation are irrelevant for herring gull populations and perhaps also for other species. The three main reasons are: *migration,* which mixes different populations in winter and exposes them to mortality factors independent of their breeding densities; *dispersal* between gulleries, which defeats any intrinsic mechanism to regulate the size of the gulleries; *communication* between different nuclei of the populations, which makes the behavior of a system encompassing a number of nuclei different from that of individual nuclei acting entirely alone. If these conclusions apply to other species, then generalized models of population regulation should incorporate the three factors.

CONCLUSION

While the myth of the balance of nature has satisfied most people, it doesn't seem to apply to the animals that biologists have studied in the field. They found *their* species to be enjoying meteoric growth or suffering catastrophic decline. Maybe the populations that are changing draw a naturalist's attention, but it is likely that these swings in local numbers are common in nature. Rather than being regulated by intrinsic factors related to density, the movement of floaters between local populations that I have described better explains the real conditions under which animals operate.

The ability of most organisms to produce a surplus of young when conditions are right has profound effects at a number of levels. Seemingly catastrophic declines in a population over several seasons may be mitigated by rapid increases in subsequent years. The production of large numbers of young by members of one species may also have effects on other organisms, as we will see in the next chapter.

SOURCES

Andrewartha, H. G., and L. C. Birch. 1954. *The Distribution and Abundance of Animals*. Chicago: University of Chicago Press.

Christian, J. J., and D. E. Davis. 1964. Endocrines, Behavior, and Population. *Science* 146:1550–60.

Clutton-Brock, T H., F. E. Guinness, and S. D. Albon. 1982. *Red Deer Behavior and Ecology of Two Sexes*. Chicago: University of Chicago Press.

Dhondt, A. A. 1971. The Regulation of Numbers in Belgian Populations of Great Tits. *Proc. Adv. Study Inst. Dynamics Numbers Popul.* (1970). 532–47.

Dhondt, A. A., and J. Huble. 1968. Fledging-Date and Sex in Relation to Dispersal in Young Great Tits. *Bird Study* 15:127–34.

Drury, W. H. 1973. Population Changes in Seabirds in New England. Part 1. *Bird-Banding* 44 (4): 267–313.

———. 1974. Population Changes in Seabirds in New England. Part 2. *Bird-Banding* 45 (1): 1–15.

Drury, W. H., and I. C. T. Nisbet. 1972. The Importance of Movements in the Biology of Herring Gulls in New England. In *Population Ecology of Migratory Birds*, 173–212. U.S. Department of the Interior Wildlife Research Report 2.

Drury, W. H., C. Ramsdell, and J. B. French. 1980. *Ecological Studies in the Bering Strait Region*. Environ. Assess. Alaskan Cont. Shelf, Final Rep. Prin. Invest. Vol. 11. BLM/NOAA OCSEAP, Boulder, Colo.

Gaston, A. J., and D. N. Nettleship. 1981. *The Thick-Billed Murres of Prince Leopold Island*. Ottawa: Canadian Wildlife Service.

Gross, A. O. 1951. The Herring Gull-Cormorant Control Project. In *Proceedings of the Xth International Ornithological Congress, Uppsala, Sweden, June 1950*, 532–536. Uppsala, Sweden: Almquist and Wiksell.

Heubeck, M. 1988. Shetland's Seabirds in Dire Straits. *BTO News* no. 158: 1–2.

Kadlec, J. A., and W. H. Drury. 1968a. Aerial Estimation of the Size of Gull Breeding Colonies. *Journal of Wildlife Management* 32:287–93.

———. 1968b. Structure of the New England Herring Gull Population. *Ecology* 49 (4): 644–76.

Kadlec, J. A., W. H. Drury, and D. K. Onion. 1969. Growth and Mortality of Herring Gull Chicks. *Bird-Banding* 40 (3): 222–33.

Kluyver, H. N. 1951. The Population Ecology of the Great Tit, *Parus M. Major* L. *Ardea* 38:99–135.

———. 1966. Regulation of a Bird Population. *Ostrich* 38 (suppl.) 6: 389–96.

Kluyver, H. N., and L. Tinbergen. 1953. Territory and the Regulation of Density in Titmice. *Archs. Neerl. Zool.* 10:265–89.

Lack, D. 1966. *Population Studies of Birds*. Oxford: Clarendon Press.

Nettleship, D. N., T. R. Birkhead, and A. J. Gaston. 1979. *Reproductive Failure among Arctic Seabirds Associated with Unusual Ice Conditions in Lancaster Sound, 1978.* Seabird Research Unit, Canadian Wildlife Service, Environment Canada, Report No. 77.

Nicholson, A. J. 1933. The Balance of Animal Populations. *Journal of Animal Ecology* 2 (1): 132–78.

Nisbet, I. C. T., and W. H. Drury. 1972. Post-fledging Survival in Herring Gulls in Relation to Brood-Size and Date of Hatching. *Bird-Banding* 43 (3): 161–72.

Perrins, C. M. 1965. Population Fluctuations and Clutch-Size in the Great Tit, *Parus Major* L. *Journal of Animal Ecology* 34:601–47.

———. 1967. The Effect of Beech Crops on Great Tit Populations and Movements. *British Birds* 60:419–32.

———. 1979. *British Tits.* Glasgow: William Collins Sons.

Slate, V., and M. Slate. Unpublished data, conversation with W. H. Drury, 1972; confirmed by J. G. T. Anderson, 1996.

Smith, S. M. 1985. The Tiniest Established Permanent Floater Crap Game in the Northeast. *Natural History* (March): 43–46.

Solomon, M. E. 1957. Dynamics of Insect Populations. *Annual Review of Entomology* 2:121–42.

Springer, A. M., E. C. Murphy, D. G. Roseneau, and M. I. Springer. 1985. *Population Status, Reproductive Ecology, and Trophic Relationships of Seabirds in Northwestern Alaska.* U.S. Dep. Commer., NOAA, OCSEAP Final Rep. 30: 127–242.

Svardson, G. 1949. Competition and Habitat Selection in Birds. *Oikos* 1:157–74.

Tinbergen, N. 1953. *The Herring Gull's World.* London: Collins Clear-Type Press.

Wynne-Edwards, V. C. 1962. *Animal Dispersion in Relation to Social Behaviour.* Edinburgh: Oliver and Boyd.

10 Habitats and Competition

Biologists and environmentalists who believe in the general existence of stable and internally organized communities that are consistent over time and space and that maintain internal homeostasis need some additional theory for completeness:

1. They need a theory to explain inconsistencies—changes in space and time. The theory of succession supplies this.
2. They need a theory to explain how populations maintain their stability to avoid fluctuations that could destabilize the community homeostasis. The theory of density-dependent population regulation supplies this.
3. They need a theory to explain how species maintain their membership in stable communities and exclude less well-adjusted species. The theory of competition and competitive exclusion supplies this.

I have discussed the theory of succession at some length in the earlier chapters of this book. In the previous chapter I questioned the general validity of density-dependent population regulation. Now I will talk about habitats in relation to ideas about competition.

FROM HABITAT TO NICHE

A habitat is a place to live, a place that is recognized and that provides the necessities of life. I discussed habitats of plants in earlier chapters while painting a *pochade* of landscapes to show how different combinations of plant species assemble on physical topography and to emphasize the ephemeral nature of these scenes. In this chapter I will address how birds use their habitats and how people have perceived the match between bird species and habitat. I will contrast my own views with the perceptions of those who subscribe to stereotyped relations and the fixed habitat units that many have called niches.

Kendeigh, in writing about the breeding warblers of a reserve in central New York, evidently accepted the assumption that habitats are fixed and their species composition remains consistent. He said, "Behavior patterns have seemingly evolved to stabilize these adjustments and maintain the species' position within the community from generation to generation with the greatest economy and efficiency of effort" (1945).

Similar ideas were then current among zoogeographers: for example, the idea of an Old World fauna, a South American fauna, an Ethiopian fauna, and so on. But, as Ernst Mayr and others made clear, species whose nearest relatives may be in South America may have very different geographic ranges in North America. The fauna does not have an identity, but species develop in certain geographic areas and then expand their ranges for differing distances in differing directions into other geographic areas.

In the late 1940s Grinnell's original identification of habitat and niche as a place to live was replaced by the "more sophisticated" idea of niche as a profession. With this replacement, focus on competition and resource partitioning redirected attention away from how birds occupy, perceive, and use their habitats in their daily lives. Tansley's representation of "mature vegetation" as stable and relatively homogenous reinforced many of the assumptions being made by the ornithologists. In addition, this sort of assumption made the formal representation of competition theory easier.

COMPETITION

Competition has been central to ecological theory ever since Darwin: "We can dimly see why the competition should be most severe between allied forms, which fill nearly the same place in the economy of nature; but probably in no one case could we precisely say why one species has been victorious over another in the great battle of life" (Darwin 1859).

Interestingly, the two founders of evolutionary theory disagreed on the role of competition. Ernst Mayr told me that while Darwin and Wallace agreed on the action of natural selection bringing out traits that allow populations to be adapted to local conditions, they differed on the concept of competition. When their joint work came out in 1858, they spoke of descent with modification. It was only later that the nineteenth-century belief in progress brought with it the Victorian idea of competition. Darwin became an advocate of it but Wallace did not emphasize competition.

The word *competition* has many different and often contradictory and confusing meanings. For some its meaning is as broad as Darwin's concept of natural selection: that which allows some individuals to survive better than others. For others, the meaning is restricted to two different individuals seeking a single item they need. The difference between death and survival is not specified. Birch wrote a helpful article in which he pointed out how diverse the meanings of the word have been: "Competition between animals occurs when a number of animals (of the same or of different species) utilize common resources the supply of which is short; or if the resources are not in short supply competition occurs when the animals seeking that resource nevertheless harm one or another in the process" (1957).

Darwin's concept of competition is important in the general idea of progress that pervades evolution as a process. As Stephen Gould reported:

Darwin did not locate the source of progress in the basic mechanics of natural selection itself—for he recognized natural selection as a theory of local adaptation only, not a statement about general advance. He justified progress with another argument about nature embodied in his favorite metaphor of the wedge. Nature is chock-full of species (like a surface covered with wedges) all struggling for a bit of limited space. New species usually win an address by driving out others in overt competition (a process that Darwin often described in his notebooks as "wedging"). This constant battle and conquest provides a rationale for progress, since victors, on average, may secure their success by general superiority in design. (1989)

RESOURCE PARTITIONING IN WARBLERS

Lack laid the foundation for much of the ecological analysis performed in the 1950s and 1960s with his work on a variety of bird species. Unfortunately, he rested his work on several important but incorrect assumptions about the world within which birds lived. Lack assumed habitats to be homogeneous, and that populations would saturate these habitats and then remain stable. These assumptions led inevitably to much broader assumptions about habitats by other researchers. The peak of consensus on stereotypy (consistency and homogeneity) was reached in the typological or essentialist treatment given bird habitats by MacArthur, Hutchinson, and their students and colleagues.

MacArthur probably began the "ecological renaissance" of model-building that lasted through the 1970s (1958). He started with assumptions of equilibrium conditions, density-dependent population regulation, competition, and strict niche segregation, and developed a topological model—a sculpture of space—for a niche. In this he implied stereotypy in both vegetation and behavior of the birds.

In his paper on warblers, MacArthur sought to reduce the differences in behavior of the several species to frequencies and directions of movements —the bare essentials (1958). He noticed that each moved about in the trees differently. To establish the statistical validity of the differences, he counted for each the place at first sighting, the number of times, and the total distance a bird moved: up-and-down, tangentially around the tree, and in toward the trunk or out to the branch tips.

MacArthur, MacArthur, and Preer proposed a system to "identify" the number of layers of vegetation (1962). The measurement, which MacArthur called foliage height diversity, provides a Newtonian order to the mechanism of bird habitat selection. The three reported:

> (1) A fairly accurate census of breeding birds can be predicted from measurements of the amounts of foliage in three horizontal layers. The abundance of each species is roughly determined by the number of patches of vegetation whose foliage profile is acceptable to that species. This suggests that many species are rare only because their chosen foliage profile is rare.
> (2) The main reason one habitat supports more bird species than another is that the first has a greater internal variation in vegetation profile (that is, greater variety of different kinds of patches). A second reason is of course that a forest with vegetation at many heights above the ground will simultaneously support ground dwellers, shrub dwellers and canopy dwellers. With a few exceptions, the variety of plant species has no direct effect on the diversity of bird species. (1962)

MacArthur's later work suggests that he, like most other American ecologists of the time, was thinking in terms of typologically perceived species and typologically perceived plant communities. He chose "typical habitat" and implied that this "niche" was subject to a Newtonian formulation—as is also evident in Hutchinson's "N-dimensional hypervolume"—which serves as a fixed backdrop for the players on his "ecological stage" (1957). This sort of formulation results in deterministic thinking about interactions perceived to be consistent over time among the same species within that "community." As I've said before, if one assumes a homogeneous and consistent habitat as the stage for species interactions, one is led to very different conclusions than if one assumes an inconsistent and changing vegetation background for the action.

More recent work by Morse further examined the nature of niche partitioning in three species of wood warblers (1971): the northern parula, yellow-rumped (myrtle), and black-throated green. Individuals of these species nested on small islands near Morse's primary study site. Morse found that he could obtain different combinations of the three species by looking at islands of different sizes. He found that "when only one species was present, it was always the Parula Warbler; when two species were present, they were always the Parula and Myrtle Warblers. Black-throated Green Warblers occurred only in the presence of the latter two species" (1971).

Should parula and yellow-rumped warblers be considered pioneer-early successional species? On the island where we have a camp, these two occupy the "climax" white and red spruce forest and are joined by black-throated green and blackburnian warblers. If I believed in succession theory, I would not expect to find a socially dominant species occurring only in the presence of the other, pioneer species. Classical niche-competition theory suggests that socially dominant species should exclude others, not regularly coexist with them.

It seems to me that parula warblers are well suited to small islands because they use the hanging clusters of the epiphyte old man's beard—which is abundant on small islands and in patches on large islands—as nest sites. For parula warblers, size of the island and presence of neighbors may be irrelevant. As to yellow-rumped warblers, MacArthur's diagrams show that these birds spend a lot of time in lower branches. The white spruce trees growing on the outside of the islands are "open grown"; that is, their lower branches flourish and provide a lavish feeding area for a species willing and able to work the lower branches. Their habitat needs design these two warblers for small islands as well as for patches of larger ones.

Movements of populations in and out are a perennial feature of several bird species on the islands. One of the most conspicuous nomads is the white-winged crossbill, which moves around searching for bumper crops of spruce cones. The birds may arrive in late July and fill the treetops with noisy singers, or they may arrive in February. They extemporize their movements and the timing of their breeding to coincide with the period when the trees of the forest are draped with cones. They are certainly not tied to any one place.

White spruce, pen-and-ink botanical
drawing, 1941.

AN ALTERNATIVE VIEW

The focus on abstract ideas of resource partitioning has distracted attention away from questions of why birds occupy so many different habitats
during migration and winter, why their habitats are often much broader
outside the breeding season, and the fact that many potential competitors
share habitats during nonbreeding periods, often joining in feeding flocks.
These behaviors scarcely suggest competition for resources and avoidance
of competitors.

Field biologists did not like the idea of fixed niches nor did they feel
comfortable with the assigning of species to such narrow stereotypes.
Darlington, for example, summarized how the idea of "niche" could be
perceived as being defined by the organism not the habitat, and be perceived as being continuously changing. Of niches, he wrote: "They are not
previously existing pigeonholes with boundaries (the concept of niche
boundaries is mathematical rather than biological), but are made and
continuously modified by the organisms that occupy them; their limits are

largely determined by competition; and the whole complex ecologic structure that results is flexible and capable of evolving both in detail and as a
whole" (1980).

Other studies of the warblers of the northeast continued to produce interesting inconsistencies in their habitat usage. Morse found that in four
species of wood warblers in Maine, males forage higher in trees than do
females (1968).

This effect could be "to avoid competition" between males and females for food, or it could reflect that the males have different "jobs" to
do. Males defend their territories by making themselves conspicuous by
singing and displaying. These activities are more effective over a larger
area if done from a higher perch. So it may be that males occupy a less
ideal habitat for feeding in order to maintain their territory; at that point
territory defense is more important than feeding efficiency. There may be
other influences, too.

Morse goes on to comment on how variable the habitats of black-
throated green warblers are from place to place (1989). These differences
suggest that the neat categories into which MacArthur put the several
species of coexisting wood warblers do not represent a sample of reality
so much as reflect his choice of samples, the restriction of samples to one
small locality, and the interpretation he chose to put on them. To this extent the treatment reinforced his search in nature for patterns stripped to
their bare essentials.

Collins focused his attention on the geographic variability in habitat
(1983). He commented that the plant communities referred to in many
studies had rarely been analyzed to determine what structural features of
the habitat were stable or variable at different sites within the range of the
species under examination. His samples indicated that the habitat of the
black-throated green warbler differs in three-dimensional structure as
well as in plant-species composition. The structural differences consist primarily of gradients from tall to shorter canopies, large to smaller trees, and
coniferous to deciduous forests.

Collins found black-throated green warblers in five types of plant community, two of rather local distribution: pine forests at Itasca State Park,
Minnesota, and spruce-arborvitae on Mount Desert Island, Maine; and

three of wider distribution: balsam fir, birch-maple-beech, and spruce-fir-deciduous forests. His report reflects a change in approach away from typological perception of vegetation as bird habitat. I like it because it tries to identify what structures within vegetation stimulate settling by this species.

His data indicate that these structures differ in different places. It suggests that in the course of history, factors acting on the members of a population returning to a geographical location influenced the membership's behavior and physical makeup. As population composition or proportions of species within their habitats change constantly, such emancipation will allow the species to persist in the face of environmental shifts.

While our perception of "patterns" is constantly changing both in time and space—as we age and move about the landscape—some associations do become apparent, and these may be important to the naturalist. From such correlations of bird species with evident patterns in the form of the vegetation came suggestions that the major climatic vegetation types had their "own" fauna—hence, the still-used terms such as "Hudsonian," "Canadian," and "Carolinian" zone species.

While these associations may "work" for the human observer, it is important to realize that birds and other animals may be making use of a different, if periodically correlated, set of habitat markers. These may be very simple symbols indeed, rather than the fine shades of meaning favored by human taxonomists. It makes good sense that animals respond to symbols to identify suitable habitats. They are not reasoning creatures, and they do not "know what is good for them." This can cause trouble: if the "correct symbol" occurs in the wrong place, birds may repeatedly go through the right motions without a useful outcome.

One spring weekend at Ernst Mayr's home in southern New Hampshire, we watched tree swallows prospecting for nests. One unfortunate male spent the weekend responding to a misleading symbol. The Mayrs had put up many new nest boxes made of tan, new wood on stakes in their meadow. Tree swallows were fluttering over the bushes and up to these boxes, hovering while chattering noisily. Males perched in the hole in each box, spreading their tails. Clearly the local tree swallows had learned to recognize squared, tan boards as signals of a prospective nest site. Earlier that year the Mayrs had had some of the shingles on their red-shingled house replaced. The isolated new tan shingles looked like the sides of the boxes in the meadow. One persistent male went through his displays in front of a single tan shingle. The bird had been confronted with a symbol it recognized, but the message was false.

Other signals may be more useful and may in fact involve the presence or absence of other species of birds, rather than vegetative or topological components of "habitat." Several Scandinavian biologists have pointed out the importance of groups of gulls already in residence on an island as a "signal" to ducks and grebes prospecting for nesting territories (Koskimies 1957). The events of the 1960s suggest that nesting herring gulls serve this "ethological signal function" not only for other herring gulls but also for colonizing double-crested cormorants, eiders, and great black-backed gulls (von Haartman 1937 and Raitasuo 1953, in Koskimies 1957). The long-term "survival value" of this response is that the presence of a busy colony indicates that the occupied island is free of predation by mammals (Olsoni 1928 and Swanberg 1932, in Koskimies 1957).

CHANGES OF RANGES OF SPECIES IN NEW ENGLAND

As we saw earlier, M. B. Davis and others have shown that the species we include in a single forest type have different rates of migration and thus have shared their habitats with very different neighbors as time has passed and climates have shifted (1980). This effect extends down to the present. Changes in species ranges have continued over the last hundred years and have been documented in northern Europe and the northeastern United States and Atlantic Canada.

During the late 1950s, tufted titmice appeared at feeders in the suburbs west of Boston. During the same years, the Carolina wrens, tufted titmice, and cardinals that had sung in New York state around Albany but not farther east, moved into the Connecticut Valley and eastward, accompanying the range extension of mockingbirds. Now one can see occasional mockingbirds, cardinals, tufted titmice, and even turkey vultures in central coastal Maine. Many people have commented on this northward spread of southern birds as if it were further evidence of a climate change, either a recovery from "the Little Ice Age" or a consequence of global heating from the greenhouse effect. Few note or remark on the simultaneous expansion southward of some "northern species," like hermit thrushes, northern waterthrushes, white-throated sparrows, and dark-eyed juncos that now nest in mixed woods in southern New England. In fact, an earlier arrival, the evening grosbeak, is a bird of northern forests.

In the late 1930s a flock of evening grosbeaks, a species that had not been seen in New England since 1888, appeared at a winter feeder in Georgetown, north of Boston. The visitation of these large birds who consumed great amounts of expensive sunflower seeds daily threatened to pauperize those who maintained the feeders, as many homeowners now know. The first invasion of any size came to Virginia in 1945–46; then a much larger invasion that produced a conspicuous influx in the Connecticut River Valley in Connecticut, Massachusetts, and Vermont, as well as Pennsylvania and Virginia, came in the 1950s (Fast 1962; Shaub 1963). The birds have persisted and spread, having begun to nest in Atlantic Canada in the 1950s (Shaub 1960; Parks and Parks 1963).

[Editor's note: This expansion of evening grosbeaks coincided with outbreaks of the spruce budworm, a food source for these birds (Nisbet, pers. com.). This is further evidence of opportunistic behavior and the importance of mobility.]

What effect does the arrival of these "pseudo-exotic" species have on "the natives?" We happened to be there, interested and watching, when the tufted titmice moved into eastern Massachusetts. We knew of their impending arrival and made a series of observations of black-capped chickadees' use of habitat before the titmice arrived. Despite our best efforts, we could find no indication that the presence of tufted titmice displaced the already resident black-capped chickadees. The titmice moved into the taller hardwood groves around the older houses, while the chickadees stayed preferentially in the "second growth" oaks and birches containing a generous admixture of pines. This is interesting, because standard competition theory, according to Elton and Lack, suggests that it was reasonable to assume a priori that the titmice were excluded from the northern area by competition.

Competition has become a necessary truth, generally accepted as the major explanation for nonoverlap in habitats and niches in "equilibrium communities," as summarized by Whittaker: "(1) If two species occupy the same niche in the same stable community, one will become extinct. (2) No two species observed in a stable community are direct competitors limited by the same resources; the species differ in niche in ways that reduce competition between them. (3) The community is a system of interacting, niche-differentiated species populations that tend to complement one another, rather than directly competing, in their uses of the community's space, time, resources, and possible kinds of interactions" (1975).

These general statements summarize what has been abstracted as a higher-level generalization from observations that related species have different habitats, especially when they occur together. The complex of theory, containing both the mechanisms and categorization of differences, is generally referred to as resource partitioning. These efforts attempt to generalize the consistent patterns into which habitat differences can be categorized, as in Roughgarden: "It would be a clear disservice to students in ecology today if textbooks failed to mention: (1) the phenomenon of resource partitioning; (2) the "checkerboard" distribution pattern of species of the same body size; (3) the regularity of the degree of difference in body size among the species who do coexist; and (4) that a possible explanation for these facts is provided by competition theory" (1983).

A key criticism of the idea that competition is centrally important in the dynamics of natural populations relates to the apparent assumption that coevolution is involved. Have the species involved maintained consistent relations with each other and evolved adaptations that permit coexistence? How can we know whether one has excluded another? Given our knowledge of the very real changes in many habitats in even the recent past, I think that the assumptions needed by advocates of competition as a general structuring force invite criticism, especially those assumptions involving environmental consistency over time and space and those with deterministic outcomes. There are plenty of situations in which competition between individuals is obvious (and usually the competing species continue to coexist), but the selective effect of this usually short-term and localized competition is not demonstrated for most situations in which "the ghost of competition past" is assumed (Connell 1980).

None of this means that competition should be ignored as a potential factor in interspecific interactions. It does mean however that competition requires experimental confirmation, that it cannot be taken as an article of faith. There are situations in which limited, stark alternatives are all that are available to individuals, and so stark contrasts in adaptations and "preadaptation" for habitat segregation is to be expected. But such situations do not mean that the segregations had to have occurred after species came together as a result of coevolution.

Rather than embrace the tidiness of theory, I have returned to more direct investigation because my own assumptions suggest that competition plays a much less central role in habitat segregation or "resource partitioning" of populations or species even though it may play an important part in the life of an individual. I assume that populations are not stable and do not saturate habitats. Rather than specializing on a narrow band of resources, each species occupies a diversity of habitats, and habitats themselves are conspicuously heterogeneous. We must appreciate as well that during most of a species' history nearly all habitats differed greatly from what we see today, in part as a result of the impact of environmental events such as ice ages. These assumptions set a very different stage on which species interact with their habitat.

SOURCES

Birch, L. C. 1957. The Meanings of Competition. *American Naturalist* 91:5–18.

Collins, S. L. 1983. Geographic Variation in Habitat Structure of the Black-Throated Green Warbler (*Dendroica Virens*). *Auk* 100:382–89.

Colquhoun, M. K., and A. Morely. 1943. Vertical Zonation in Woodland Bird Communities. *Journal of Animal Ecology* 12:75–118.

Connell, J. H. 1980. Diversity and the Coevolution of Competitors, or the Ghost of Competition Past. *Oikos* 35:131–38.

Darlington, P. 1980. *Evolution for Naturalists.* New York: John Wiley and Sons.

Darwin, C. 1859. *On the Origin of Species by Means of Natural Selection.* London: Watts.

Darwin, C., and A. R. Wallace. 1858. *Evolution by Natural Selection.* New York: Johnson Reprint Corporation.

Davis, M. B. 1980. Quaternary History and the Stability of Forest Communities. In *Forest Succession*, edited by D. C. West, H. H. Shugart, and D. B. Botkin, 132–53. New York: Springer-Verlag.

Fast, A. H. 1962. The Evening Grosbeaks in Northern Virginia. *Bird-Banding* 33 (4): 181–91.

Gould, S. J. 1989. Tires to Sandals. *Natural History* (April): 8–15.

Hartley, P. H. T. 1953. An Ecological Study of the Feeding Habits of the English Titmice. *Journal of Animal Ecology* 22:261–88.

Hutchinson, G. E. 1957. Concluding Remarks. *Cold Spring Harbor Symposia on Quantitative Biology.* 22:415–27.

Kendeigh, S. C. 1945. Community Selection by Birds on the Helderberg Plateau of New York. *Auk* 62:418–36.

Koskimies, J. 1957. Terns and Gulls as Features of Habitat Recognition for Birds Nesting on Their Colonies. *Ornis Fennica* 34:1–6.

MacArthur, R. H. 1958. Population Ecology of Some Warblers of Northeastern Coniferous Forests. *Ecology* 39 (4): 599–619.

MacArthur, R. H., J. W. MacArthur, and J. Preer. 1962. On Bird Species Diversity. II: Prediction of Bird Census from Habitat Measurements. *American Naturalist* 96:167–74.

Morse, D. H. 1968. A Quantitative Study of Foraging of Male and Female Spruce-Woods Warblers. *Ecology* 49 (4): 779–84.

———. 1971. The Foraging of Warblers Isolated on Small Islands. *Ecology* 52 (2): 216–28.

———. 1980. *Behavioral Mechanisms in Ecology.* Cambridge: Harvard University Press.

———. 1989. *American Warblers.* Cambridge: Harvard University Press.

Parks, G. H., and H. C. Parks. 1963. Some Notes on a Trip to an Evening Grosbeak Nesting Area. *Bird-Banding* 34 (1): 22–29.

Roughgarden, J. 1983. Competition and Theory in Community Ecology. *American Naturalist* 122 (5): 583–601.

Shaub, B. M. 1960. The Evening Grosbeak Incursion in the Northeast Winter of 1957–8. *Bird-Banding* 31 (3): 140–50.

Shaub, M. S. 1963. Evening Grosbeak Winter Incursions—1958–59; 1959–60; 1960–61. *Bird-Banding* 34 (1): 1–21.

Whittaker, R. H. 1975. *Communities and Ecosystems.* 2nd ed. New York: Macmillan.

11 Predation, Frugivory, and Pollination

STRONG FORCES OPERATING BETWEEN SPECIES

Much of the interest in natural selection has focused on interactions between both individuals and species. It is important to realize that not all interactions are equal in terms of their likely effects on survival and reproduction. I suggested in the previous chapter that competition, whether within or between species, is likely to have a relatively weak selective effect under most circumstances. By way of contrast, there are certain interactions that may have a much more immediate effect on the survival of individuals, and hence can serve as strong forces in natural selection. Just how strong these forces actually are may vary from place to place and from species to species.

PREDATION

Predation can have significant consequences on prey populations because predators choose between individuals, and if their choices are based on specific traits in the prey, the predator acts as a key element of natural selection. Predators have attracted the positive and negative notice of many people, and they are a popular feature in many folk traditions. Some of these traditions form the basic assumptions that many of us make when we first examine the concept of predation. The Greek historian Herodotus stated that edible animals breed rapidly, and that predators keep their numbers under control, a notion repeated two thousand years later by Linnaeus. This idea persists in the policies of many environmentalists and state fish and game departments, that "natural enemies" regulate and maintain the health of the prey population and thereby "nature's balance."

My animal ecology teacher emphasized integrative explanations of the distribution of organisms in preference to "purely descriptive" natural history. Thus we heard a great deal about the roles of predators in communities: that of regulating population density and determining species composition; and we heard relatively little about a predator's actual activities: where they worked and what they caught.

While my teacher recognized field-based studies as being valuable, he clearly preferred Gause's experiments on the effects of *Paramecium* predation on *Didymium,* carried out in laboratory aquariums (1934). He emphasized the importance of representing the results in mathematical form, the logistic equation that we discussed in chapter 9. He cited the elegant derivation of the Lotka-Volterra equation that seemed to describe the predator-prey interactions between red fox, Canada lynx, and snowshoe hares. The data that most interested him gave validity to the mathematical model of linked predator-prey fluctuations in natural systems.

Much of the same thinking still appears in recent texts on biology and ecology because several generations of ecologists have accepted the validity of the Lotka-Volterra equation for populations at an assumed equilibrium. This logic then creates a necessary role for natural enemies in the "balance of nature." In contrast, I believe that predation often amounts to little more than the dissipation of excess.

While examples of tight coupling between pairs or even groups of species have been identified, we need to ask whether there really is a degree of "ecological equivalence" among species in terms of their impact on what some researchers have identified as communities or ecosystems.

There is a tendency in systems thinking to lump groups together under simplistic headings such as "predators" and "prey," without paying adequate attention to the very real differences that the biology of members of each group imposes on their activities. Predators are not simply optimal killing and eating machines, and prey do not submit willingly to their role as food. Detailed studies of bird predation and predation in other groups of animals suggest that limitations imposed by relatively fixed-action patterns produce effects that contradict the expectations of those seeking simplistic demonstration models of population regulation.

Predatory Hawks and Vulnerable Songbirds

Niko Tinbergen suggested how the interactions between predators and prey might reflect chance matches between a hunter's search pattern and the hunted's vulnerability (1951). He proposed that a peregrine falcon's hunting behavior is made up of a chain of events in which hunting behavior starts as a very general class of activities and is progressively focused onto an individual victim. He suggested that the hunt starts with the hawk feeling an unlocalized uneasiness that stimulates the bird to fly in haphazard directions until it sees some potential prey. Sighting prey shifts the "program" and the falcon makes a test "run" at the prey.

This first run may give both the raptor and its prey information about each other. A dallying raptor circles and flies steadily or does not follow up on a momentary pass, and thus represents only a minor threat. The focused attention of a raptor rapidly closing in, in contrast, is a clear indicator to prey that it is in danger. Whatever the motivation of a raptor making a pass, the potential victims usually take evasive actions. It is possible that some of the evasive actions are signals to the raptor that its approach has been detected and that the potential victims are alert and are not worth further pursuit.

It is useful to think of hunting actions as species-specific patterns of relatively fixed actions among most birds, whether they are hawks, owls, songbirds, or waterfowl. Luuk Tinbergen described the hunting of sparrow hawks as stereotypical (1946). The birds' individuality is expressed in applying a limited repertory of behaviors to different landscapes and in concentrating on certain hunting routes where each individual had been successful.

Luuk Tinbergen showed that habitat preferences and species-specific behavior made some species of prey much more vulnerable than other species equally abundant and widespread. House sparrows and tree sparrows were most vulnerable, chaffinches and great tits were somewhat less so. House sparrows suffered because they were found primarily in those habitats where the hawks did most of their hunting, and because they often left cover, were always busily moving about, and lived socially throughout the year.

The male sparrow hawk does nearly all the hunting during incubation and while the young are small, and in these periods the percentage of woodland birds taken is high. As the nestlings grow, the proportion of woodland birds used by the hawks drops off as the male hunts more in the villages and along farmyards. The parent's hunting routes include more fields after the chicks fledge, as suggested by increases in the catch of starlings and yellow buntings. In general, the few species like house sparrows, whose habitat and behavior match the hawk's hunting techniques, are exploited and the hawks are an important part of the mortality of these species.

Other detailed studies of bird predation suggest limitation set by relatively fixed action patterns, yielding results that contradict the expectations of how predation "regulates animal populations" as the Lotka-Volterra model proposes.

Ranges of Prey Taken

One idea in widespread circulation is that predators take primarily old or sick prey, rather than simply select whatever prey can be surprised or happens to be in the right place at the right time. In his studies of raptor hunting, Rudebeck tried to establish whether individual prey are taken randomly or some individuals were more vulnerable (1950). It became evident that the differences between individuals within a prey species were more important than were the differences between species.

Rudebeck defined a "hunt" as an attempt to kill or seize a quarry once it had been selected. Rudebeck counted twenty-three catches out of 190 hunts. Of these only one victim was evidently injured or in some way at a physical disadvantage prior to the hawk's attack, but five of the twenty-three showed atypical behavior that put them in an exposed position.

It seems likely that even large raptors have difficulty catching large prey, and that they must keep cruising until they find some individual that is down on its luck. Smaller prey on the other hand may be handled relatively easily, especially if it is surprised. The general idea that predators cull the injured and sick may be in part a phenomenon of what catches the human eye.

In a similar fashion, if we examine feeding behavior of seabirds, we see what is probably an inevitable consequence of the highly variable environment they inhabit. The specific search image, or picture of what to look for and where, of kittiwakes seems to be simple. They feed on any small, shiny, edible items floating near the surface. They will take shrimplike euphausids and amphipods, but they greatly prefer to plunge for silvery "bait fish" flashing in shoals near the surface. This relaxed attitude about what is suitable food allows kittiwakes to use markedly different prey species in different parts of their range (Belopol'skii 1961; Baird and Moe 1978; Drury et al. 1980; Salomonsen 1950; Springer et al. 1984). They take whatever small, silvery-sided shoaling fish is available at the surface and will specialize on any abundant prey that is present and accessible. We also see this lack of prey-specificity in terns of the northeastern United States.

A population of predators can eliminate a species from its habitat if the prey have inadequate cover and if their activities make them evident to the predator. On the other hand, predators may have a negligible effect where the prey's cover is abundant. If the predator does affect the prey, this impact can take two major forms. The first is that a predatory species may have strong effects on the numbers of some species in its "community" while having only weak effects on many others. The second is that even while predators may have little impact on the size of the prey population, they have a marked effect on the characteristics of the prey. Predation is a potent selective force, because predators select from among populations of prey with variable characteristics.

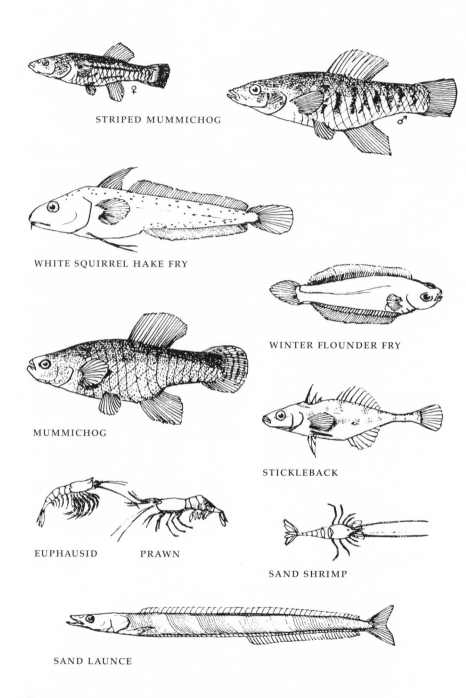

STRIPED MUMMICHOG

WHITE SQUIRREL HAKE FRY

WINTER FLOUNDER FRY

MUMMICHOG

STICKLEBACK

EUPHAUSID PRAWN

SAND SHRIMP

SAND LAUNCE

Figure 3. Prey species used by terns in the northeastern United States. Like the kittiwake, terns respond to their variable environment by feeding on whichever species is present and accessible.

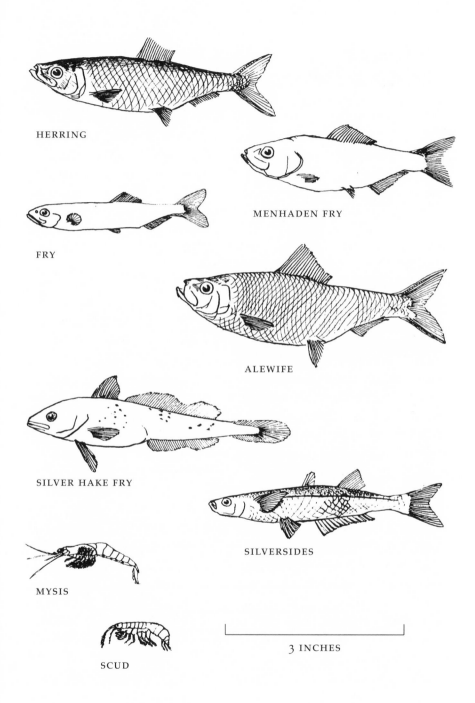

HERRING

MENHADEN FRY

FRY

ALEWIFE

SILVER HAKE FRY

SILVERSIDES

MYSIS

SCUD

3 INCHES

Bird Predation on Pest Insects in Pine Plantations

During the furor over the use of chlorinated hydrocarbon pesticides in the 1950s, many biologists were stimulated to investigate the part played by wild birds in the control of insect pests in commercial forests. Luuk Tinbergen investigated the factors influencing the effects of predation by tits, primarily the great tit, on caterpillars that defoliate trees (1960). He concentrated on measuring the populations of tits each year, the density of certain caterpillars, and the frequency of each prey species in the diet of nestling tits.

One of Luuk Tinbergen's major finds was that the frequency at which a prey species appeared in the great tit's diet depended not simply on density but also on size, conspicuousness, and a rather vague concept he called "palatability." This led to the most widely quoted and debated of Tinbergen's conclusions. He observed first that even after suitable caterpillars appeared in the birds' feeding area, they did not appear in the nestlings' food for several days; second, that suddenly the numbers of these caterpillars in the food increased abruptly; and finally that the delay varied from bird to bird but was less variable between the members of a pair. Combined, these observations led him to suggest that the birds overlook the new kind of caterpillar until they notice it by chance. Then they form a specific search image and may concentrate on it.

While most people quote his study for the concept of specific search image, I am interested in another of Luuk Tinbergen's observations. He noticed that even after discovery the percentage of each species of caterpillar in the diet did not vary directly with the frequency expected from "chance encounters." When he calculated the frequency of insect larvae in the pine foliage and adjusted this by an index of risk, he found that the birds used the four common species of caterpillars much less than expected at very low densities; that their use increased abruptly and exceeded expectation at moderate densities; and that the use declined below expectation at high densities. He explained the first as reflecting the formation of a search image. He explained the second change as indicating the birds' preference for a mixed rather than a monotonous diet. No matter how abundant, no single species of caterpillar made up more than half of the birds' diet. This is what he meant by palatability. His observations are more important than his theoretical explanations.

The fact that bird predation was most intense at moderate prey densities and declined markedly at high densities is the opposite of what Nicholson proposed for density-dependent regulation of insect populations (1933), but the late 1950s was just the wrong time to say that.

Predation by Expectation   Gibb found the feeding techniques of coal and blue tits to be similarly restricted where they fed primarily on the caterpillar infesting the cones of Scots pines in England (1958). The density of infestations was markedly patchy, and the variations in density over short distances were often spectacular. Tits might meet a larger range of intensities in small, highly infested plots than they would meet when feeding over most of the compartments in the pine plantations.

Birds would cross several areas of high and low extremes of density in the course of a day's feeding. They searched for the concealed prey by tapping on the cone scales. If the tit obtained the correct clue it dug into the scale and extracted its prey, leaving a scar on the scale's tip. If the larva matured and emerged, it left a small round hole. The scars or neat holes provide a basis for counting how many larvae were in each cone and how many were taken by tits.

The tits did not use a "random" sample of the insects available. Gibb found that tit predation on the larvae in the pine cones consisted of two parts: a rather stereotypical pattern of search and a rather fixed pattern of expectation. The birds would tap on half a dozen scales. If they found none that "felt right," they would move on to the next cone. If they were rewarded, they would tap on some more scales of the same cone while they were still being rewarded. But he found that the birds would consistently move on well before they had exhausted the possibilities available in seriously infested cones. It was as if they couldn't believe that there would be that many larvae in one cone. More probably, the odds of there being many larvae per cone were low enough that a better overall strategy was to keep moving.

Thus they averaged out the intensity of their search to match a realistic expectation—what produced the most favorable results over most of the pine plantations. The patchiness of the infestation made it rewarding for the tits to keep looking rather than become preoccupied with finding one more worm where they were. It is usually better to grab the obvious prey and move on, even in "hot spots."

Predation by Profitability   Royama studied the feeding of great tits in larch plantations in Japan and in mixed broadleaf woods in England (1970). He was interested in what drives the ways parent birds choose the prey they bring to their nestlings. The only consistent correlation Royama found was that in all broods the proportion of spiders in the diet increased for the first few days after hatching, reaching a maximum on about day five or six and then gradually decreasing.

Royama found no direct relation between the volume (biomass) of the prey in the habitat and the prey the tits brought in, but many prey species appeared in numbers as the larvae pupated, suggesting that conspicuous movements during pupation drew the tit's attention. His main conclusion was that the titmice work to improve their efficiency at bringing in food. During the early part of the breeding season they hunted mainly in oak foliage, the habitat of the most abundant prey they were bringing in. During the middle part of the season, other species predominated and some tits worked over blackthorn, hawthorn, and ash. But even then they spent a lot of time in oaks. After the middle of June, drastic changes occurred. The birds shifted to prey species living on herbaceous plants on the forest floor.

The efficiency of hunting can be expressed in terms of the number of prey that a predator can collect in a given time. The efficiency should be influenced at any given density by the predator's ability to find the species, by the handling time for each item, and by the weight of the prey species. Royama therefore introduced the concept of "profitability." Profitability is measured as the amount of prey, in terms of biomass, that the predator can collect in a given time spent hunting.

If this explanation is correct, differences in handling and commuting time may influence the order of profitability of prey species of different sizes and explain the differences in size of prey in the diet of nestlings and adults respectively.

In 1969 a party of five of us had an experience in predation by profitability while on a project to band twenty thousand gull chicks. We wanted to find every chick alive and were willing to spend a lot of time searching under rocks and bushes for concealed chicks. In a sense our objectives resembled those of the feeding tits—to find as many "prey" as possible in a limited amount of time.

While four of us plodded on in our search, each banding less than thirty chicks for the morning, one of the party banded over eighty. We noticed the difference and asked after his skill. Easy, he said: he just kept moving, banding the most conspicuous birds, neglecting time-consuming searches in crevices. I expect that when rewards slacken, tits catch more prey by moving on than by concentrated search.

"Cycles" of Predators and Prey

Various circumstances allow populations to build up far in excess of ordinary numbers. This seems to happen where there are extensive areas of continuous, favorable habitat. The sudden population flux may be caused by access to concentrated food, as in house mouse plagues, and/or locally excellent shelter once numbers are high, as when lemmings gather under persistent snowbanks in spring. These situations make it advantageous for individuals to accommodate crowding. This is especially true if parents are faced with the choice of allowing their own offspring to stay in the territory or chasing them out to face almost certain mortality. These circumstances can overwhelm ordinary patterns of spacing and are clearly visible at bird feeders, on fish gurry, and on large carcasses.

After concentration of numbers has occurred, something sudden and unusual—such as the removal of cover with melting of the snowbanks, change or gathering of the crops, emptying of a corn crib—suddenly exposes large numbers of individuals at once, and while some individuals still persist in the remaining cover, most emigrate and scatter.

If a species characteristically cycles or if cycles occur in a particular place or under certain conditions, you can explain each case, but it is dangerous to expect that those explanations will apply to other situations. Species are different, habitats are different and cycling is a special case in which mortality, instead of holding steady at about 10 to 14 percent for large animals or 25 percent for small ones, drops for several years and then goes to 90 percent or more in a short period. One should expect the events that lead to this to be capricious and not subject to a priori logic.

Pitelka, Tomich, and Triechel studied the cycles of lemmings and large bird predators at Point Barrow, Alaska (1955). The authors followed events among the lemmings and the bird predators attracted to the area, starting at the low ebb of the lemming cycle. While the lemmings were building their numbers, females reached sexual maturity early and bore large litters several times a year. After the populations became dense, they suffered a time of intense crowding in springtime as the thaw removed the snow roofs of their familiar winter tunnels. As the snowbanks shrank, the animals were forced into smaller and smaller areas; individual contacts, stress, and conflicts intensified; and eventually the animals exploded out over the tundra.

Migrating pomarine jaegers rove over the coastal tundra, "hopeful" that they will happen on a place where the lemming population is overflowing. Most years they don't. The crew at Barrow found one jaeger nest in 1950. In 1951, the crew saw groups of three or four arrive on June 6, stay until June 30, and disappear. In 1952 the crew saw jaegers on June 9 and found thirty-four nesting pairs that year—1952 was the "good year." Most pomarine jaeger nests contained two eggs; little fighting interrupted laying, and successful nests produced two young.

In 1953, pomarine jaegers arrived May 25 and settled on about 128 aggressively defended territories. The density of nests increased by four times compared to the previous year. Pitelka and his colleagues associated this with the conspicuous lemmings that were exposed as the snow melted off the tundra. The numbers of lemmings dropped soon after the birds settled, territorial squabbling among jaegers became violent and conspicuous, laying was disrupted, and many nests were destroyed or abandoned. Eggs appeared over twenty days and many were infertile. Many adults left their nests before the young were independent. The few successful nests each produced only a single young. A floating population of nonbreeders reformed flocks at the end of July. By August 8 both jaegers and lemmings were scarce. In 1954 lemmings were scarce and the jaegers arrived singly; the crew found no nests.

Snowy owls also increased at Barrow when there were several "good lemming years." Owls overwintered in 1948–49 but Pitelka and his crew found no nests. They were present in small numbers in 1949–50, and again in 1951, but there were no nests. In 1952, five or six pairs set up territories one to two miles apart and remained until the end of the season.

In 1953, ten to twelve owls were nesting on Barrow Spit the first week of June. Their numbers increased to thirty-three by the 15th, and non-breeders continued to increase all through June, suggesting that owls were being driven off the high tundra by jaegers. The owl nests were much farther apart than those of the jaegers, and some territories occupied two square miles. Pitelka and his crew found nests containing from four to nine eggs.

At one nest of satiated young, they found eighty-three lemmings that were from broods of the previous winter or before, not from the summer's crop of young. Full-grown owl fledglings weigh about 1,600 grams and eat three to four lemmings a day. At the end of July of 1953 nestling owls weighed between 77 and 545 grams, and food was already short. Of thirty-one eggs in the nests they watched, only two young birds fledged.

Pitelka and associates concluded that predators were not responsible for the downfall of the lemming population. Instead the predators delayed the crash by cutting into the population before it peaked. Populations of prey increased in spite of predation, and predation served to truncate the peak and dampen the swings, extending the time between peaks.

The timing of the explosion of lemmings out of the snowbanks and the subsequent population crash was important. In off years the lemmings might become very numerous after a winter's successful survival and reproduction, but dispersal and crash did not occur. In these circumstances, smaller numbers of jaegers and owls might be attracted, might breed, have plenty of food, and raise young. When, however, dispersal and crash occurred in early June, and many birds were attracted, the birds' food supply evaporated during incubation, creating a lot of squabbling over what little food there was. Many young jaegers and owls starved.

Overwintering lemmings are exposed in early June by snowmelt, and during a crash pregnancy rates are low or nil. Consequently, just as the predators are starting to breed, their prey fails, and by mid-July only one-tenth to one-twentieth of the mid-June food supply is available.

There is a popular conception that lemmings regulate their population through mass drowning when their numbers get too high. Aho and Kalela found that the story is more interesting than that (1966). Many individuals do become rovers and rush off to try to find someplace new, and many of these dispersing individuals may be caught in streams and washed into the ocean; but many remain at home. Thus there are at least two alternative sets of founders of the next population "high": the surviving dispersers and the individuals who remained on their natal range.

Most people have expected mass migrations of lemmings to occur in North America, because they thought that the lemming cycles occurred synchronously all across the Canadian north. But better information showed that lemming cycles are largely disassociated from each other. To this extent the waves of lemming populations mounding here and there over the north are like the waves in a tide rip or below rocky cliffs that occasionally reinforce each other to produce one great wave. Some individuals survive in places that were missed by the population boom/bust, and some survive by luck in the midst of mass mortality. Then, a year of exceptional breeding success produces a year class that may represent the bulk of the population for a dozen years. The young disperse and a wave of emigrants is propagated outward to recolonize areas that were emptied in the last crash.

Adaptations of Prey to Avoid Predation

The persistent pressure of predators searching for prey, finding some individuals and missing others, has brought out traits in the prey that confuse the predator or make it hesitate. The plumage of ground-nesting songbirds, shorebirds, and waterfowl matches its vegetation background. The match of eggshells to their background is no less elegant.

Dispersal and Concealment of Nests   I presume that wide dispersal of nests, the choice of nest site, the eggs' elegant match to the background, and the ability of newly hatched young to run, hide, and freeze motionless are all adaptations brought out by the excruciating pressures of individual predators finding some individuals and missing others. Many observations exist of the subtle adaptations by which some waterfowl counter predation.

Niko Tinbergen showed how the gull habit of carrying away eggshells after the young hatch removes objects that might attract a crow's or fox's attention (1953). The inside of the shell, being white and contrasting with the outside, is much more conspicuous than is the mottled outside. Niko Tinbergen and Kruuk showed that the habit of gulls nesting widely spaced among sand dunes is a further adaptation (N. Tinbergen, Imprekoven, and Franck 1967; Kruuk 1964, 1975). They discovered that nests closer together were more vulnerable to predation. Once a fox has found one nest, it will search diligently around it for a while, then, like Gibb's coal tits, its attention shifts and it trots off when its "expectation of finding the next nest" is not met.

Tinbergen and Kruuk reasoned that the wide spacing of gull nests on sand dunes reduces the success of prowling foxes and helps to minimize predation on any one nest. Our experiences on gull islands seemed to confirm this, and added some nice details. We found nests to be widely spaced (twenty to thirty-five meters apart) on sand dunes in the region of sandy shores in southern New England where the movement of sand spits repeatedly builds spits and cuts them into islands, where gulls nest. Foxes usually have access to these gulleries. The gulleries on rocky islands on the coast of Maine, however, are not about to be connected to the shore, and hence are less vulnerable to terrestrial predators. The gulls nesting amid piles of driftwood or small boulders usually nest within one to five meters of each other.

Schools, Flocks, and Herds   With the renaissance of Darwinian thinking, biologists' attention shifted from studies that assumed a benefit to the group as a whole to investigating individuals in flocks acting in selfish interest. If a predator is looking for a victim, each individual's odds of being taken are lower if there are several alternative victims, especially for the one in the middle when others are exposed on the margins. Krebs and Davies raised another advantage to flocking behavior: "For many preda-

Blue jay, pencil, circa 1955.

tors success depends on surprise: if the victim is alerted too soon during an attack, the predator's chance of success is low. This is true, for example, of goshawks hunting for pigeon flocks . . . : the hawks are less successful in attacks on large flocks of pigeons mainly because the birds in a large flock take to the air when the hawk is still some distance away. If each pigeon in the flock occasionally looks up to scan for a hawk, the bigger the flock the more likely it is that one bird will be alert when the hawk looms over the horizon. Once one pigeon takes off the others follow at once" (1981). Thus, by belonging to a flock, the individual can spend less time looking for predators and more time feeding.

Warning Coloration and Mimicry   Distastefulness or toxicity to herbivores is widespread among plants. Many species sequester secondary metabolic products that happen to have often violent effects on animals that consume them. If an insect can sequester the poison internally, it gains protection from its own predators. This is well illustrated by the case of monarch butterfly caterpillars made distasteful by feeding on milkweeds.

Brower showed that some monarchs contain toxic cardiac glycosides, and that a blue jay who eats a poisonous monarch becomes ill and vomits fifteen to thirty minutes later (1969). That single encounter is sufficient to make the jay avoid the butterfly thereafter.

Monarchs get their poisons from the milkweed plants on which the females lay their eggs. It is the milkweeds that evolved the glycosides that interfere with an animal's internal activities. But the larvae of monarchs have evolved the ability to feed on milkweeds without serious harm. They store the poisons in their tissues, holding them even when they metamorphose from a larva into an adult.

Brower investigated the mechanisms involved by giving monarch larvae no choice but to feed on a diet of cabbages rather than milkweeds. When he fed the adults that emerged from these larvae to jays, the jays suffered no ill effects; the jays fed butterflies reared on the milkweed always vomited a short time later. Chemical analysis of such butterflies showed that the glycosides in the milkweeds were identical to those in the butterflies.

The monarch caterpillars and butterflies also show warning colors— striking yellow, black, and tan stripes on the caterpillar and orange and black markings on the butterflies. Behavior in female monarchs varies: not all lay their eggs on milkweeds. But those that happen to lay their eggs on nontoxic food plants benefit by the lessons predators learn from bitter experience with toxic ones. Predators with experience with milkweed-reared monarchs are likely to avoid the monarch pattern on sight.

Notice that the individual who provides the lesson to the predator doesn't benefit, but its parent does. The lesson provides better survival for all the offspring of the parent that lays eggs on milkweed. In this way parents "play the lottery," expending some young to increase the odds that some will come through. The more young that parents produce, the more they can expend in the interests of the majority. If parents lay one hundred eggs, the experience of predators that tasted ten larvae would benefit the remaining ninety. So, the more conspicuous, distasteful, poison-laden caterpillars the better. This means that the odds are increased for all concerned if several distasteful species resemble each other.

Warning colors are like cryptic colors. Even subtle differences may make a predator hesitate and allow a potential victim the chance to live another day. So it is also possible for a good-tasting caterpillar to benefit by even a passing resemblance to an evil-tasting one. We should not be surprised to learn that a number of caterpillars and butterflies that are eminently edible have evolved to mimic distasteful ones. This protection by deception has been cited for the fact that a number of edible flower flies closely resemble yellow-and-black yellow jackets. The difference between four wings on the yellow jackets and two on the flower flies is pretty subtle to pick up at a quick glance.

While the notion of tightly coupled groups of predators and prey is soothing for those seeking order and consistency, the variability of habitats and populations is likely to relegate it to the realm of theory. We will have to look elsewhere for better examples of coevolutionary relationships.

POLLINATION AND FRUGIVORY

The dancelike exchange of influences between insect pollinators and their plants is one of the most highly developed expressions of how strong the forces of natural selection can be. Plants and their pollinators have co-evolved; they are *really* members of the same community. Plants and animals take turns acting as agents of natural selection when the plants need insects to carry pollen and the insect larvae need the plant's ovules for food. Yet many plants are independent of animal pollinators. Pines, which are among the oldest and considered the most primitive of the woody plants, as well as grasses, which are considered among the most advanced, both grow in extensive stands covering many acres. This growth in ranks allows them to depend on the wind to disperse their pollen.

Grant provided an excellent description of the interplay (1951). If plants need some third party to transfer pollen between two individuals, that activity must prove advantageous to the living agent carrying the pollen. So flowers have become adapted through natural selection to the characteristics of their pollinators. The structure, shape, color, odor, and so on of flowers are responses to selection by the agents that cross-pollinate them. Flowers are grouped into several broad classes according to how they are pollinated: fly flowers, beetle flowers, bee flowers, moth flowers, bird flowers, bat flowers, and wind flowers. Plants visited by unspecialized insects such as beetles and flies provide their pollinators with little except odors and take care to protect their own seeds from depredations. But other plants provide specialized insects such as bees with nectar, pollen, shelter, a landing platform, bright colors, and sweet odors.

The fact that healthy plants can produce leaves, branches, and seeds in great excess has resulted in plants expending fruits to disperse their progeny. As dispersal is the primary benefit, plants that attract several species of animals will be favored, especially if the animals come from and return to many heterogeneous places with several types of habitat. We might expect plants to do well whose fruits match the feeding methods of several species with not-too-specialized habits.

The patchy distribution of plants affects their success in dispersing their fruits, and this in turn may be affected by the fruiting of related plants in the same area. Herrera found that species with less popular fruits were better dispersed when growing in the company of species with preferred fruits (1984). This effect seems to be mediated by the dispersers having physiological need to use several different fruits, as no single kind of fruit provides a complete diet.

Herrera also became interested in historical changes in traits of fruits, presumably in response to selection by dispersers (1986). He found in the Lauraceae that traits have remained remarkably constant for long periods of time. Fossils show little change in fruit structures, and geographical evidence indicates that closely related species have very similar fruit characteristics although they diverged long ago and are now found in different habitats and dispersed by different bird species.

Yet Herrera's experience shows that the plants do respond to the general pressures that dispersers exert. Fruits ripen in each month in southern Spain. They are taken by different species in different months. Herrera found that more watery fruits were available when water was scarce, while in winter more energy-rich fruits were available (1981). These shifts suggest that plants supply berries that will be as attractive as possible to their dispersers. When they have a choice, frugivorous birds favor fruits that best fill their needs.

Fruiting plants have their own Madison Avenue effect. Plants advertise their fruits with showy colors to attract the attention of their dispersers. As you might expect, plants do not invest any more energy in their dispersers than they have to, and the plants do not have the same

Yellow lady's slipper, pen-and-ink botanical drawing, 1941.

loyalty to their dispersers that they have to their pollinators. Flowers "want" their pollinators to come back, to visit another individual of the same species. In contrast, fruiting plants "want" their dispersers to go away to different habitats.

Frugivores' habitats include many patches and many alternative multispecies combinations of both plants and dispersers, and these change across the landscapes and over time. It seems to have been prudent to remain conservative and generalized rather than to specialize.

The limitations of habitat suitable to a particular plant will limit the success of dispersed seeds to those taken by dispersers that occupy or pass through habitats within an acceptable range. Dispersers who time their movements to coincide with changes in the abundance of seeds and fruits can take advantage of resource availability. The greater the variety of potential fruits and seeds to be dispersed, and the greater the variety of dispersal mechanisms, the weaker we should expect selection to act on individuals. Flexibility and disengagement from codependency is likely to be favored in varying environments, rather than the tight coupling that has been predicted in many descriptions of assemblages of plants and animals.

CONCLUSION: CONFLICTING THEMES

In the examples that we have discussed here, we can see both the strong forces of predation, frugivory, and pollination on the lives of many individuals and the apparent consequence of these activities in terms of natural selection. The ability of organisms to produce many more young than is required for strict replacement dilutes the effectiveness of predation, both as a mechanism of "population control" and as a force in evolution. The widely held notion among many environmentalists that "predators keep the balance" is not supported by field evidence, and indeed the example of small mammals in the north suggests that the reverse may be true: predators are opportunists who boom and bust or immigrate and emigrate in response to changes in prey.

Frugivory and pollination, situations in which both groups of participants may benefit, are more likely candidates for coevolutionary relationships. In pollination it is to the advantage of the plant that a pollinator specialize in movements between members of the same species. Selection on the structure and behavior of both plant and pollinator may be expected to be intense. In frugivory, the advantage to plants of wide dispersal of their propagules may lead to selection for traits that attract an array of dispersers rather than a single species.

SOURCES

Aho, J., and O. Kalela. 1966. The Spring Migration of 1961 in the Norwegian Lemming, *Lemmus Lemmus* (L.), at Kilpisjarvi, Finnish Lapland. *Annales Zoologici Fennici* 3:53–65.

Baird, P. A., and R. A. Moe. 1978. Population Ecology and Trophic Relationships of Marine Birds at Sitkalidak Strait, Kodiak Island, 1977. In *Environmental Assessment of the Alaskan Continental Shelf.* NOAA/OCSEAP, Ann. Rep. 3:313–524.

Belopol'skii, L. O. 1961. *Ecology of Sea Birds' Colonies of the Barents Sea* (translated from Russian). Jerusalem: Israel Program for Scientific Translation.

Brower, L. P. 1969. Ecological Chemistry. *Scientific American* (February): 22–29.

Drury, W. H., C. Ramsdell, and J. B. French. 1980. *Ecological Studies in the Bering Strait Region.* Environ. Assess. Alaskan Cont. Shelf, Final Rep. Prin. Invest. Vol. 11. BLM/NOAA OCSEAP, Boulder, Colo.

Gause, G. F. 1934. *The Struggle for Existence.* New York: Dover Publications.

Gibb, J. A. 1958. Predation by Tits and Squirrels on the Eucosmid *Ernarmonia Conicolana* (Heyl.) *Journal of Animal Ecology* 27:375–96.

Grant, V. 1951. The Fertilization of Flowers. *Scientific American* 184 (6): 52–56.

Helms, C. W., and W. H. Drury. 1960. Winter and Migratory Weight and Fat: Field Studies on Some North American Buntings. *Bird-Banding* 31:1–40.

Herrera, C. M. 1981. Fruit Variation and Competition for Dispersers in Natural Populations of *Smilax Aspera.* Oikos 36:51–58.

———. 1984. A Study of Avian Frugivores, Bird-Dispersed Plants, and Their Interaction in Mediterranean Scrublands. *Ecological Monographs* 54 (1): 1–23.

———. 1986. Avian Frugivory and Seed Dispersal in Mediterranean Habitats: Regional Variation in Plant-Animal Interaction. In *Acta XIX Congressus Internationalis Ornithologici,* 509–17. Vol. 1. Ottawa, Canada: University of Ottawa Press.

Krebs, J. R., and N. B. Davies. 1981. *An Introduction to Behavioural Ecology.* Sunderland, Mass.: Sinauer Associates.

Kruuk, H. 1964. Predators and Anti-predator Behavior of the Black-Headed Gull (*Larus Ridibundus* L.) *Behavior* suppl. 11:1–129.

———. 1975. Defense against Killers. *Natural History* 75 (4): 48–55.

Nicholson, A. J. 1933. The Balance of Animal Populations. *Journal of Animal Ecology* 2 (1): 132–78.

Pitelka, F. A., P. Q. Tomich, and G. W. Triechel. 1955. Ecological Relations of Jaegers and Owls as Lemming Predators Near Barrow, Alaska. *Ecological Monographs* 23 (3): 85–117.

Royama, T. 1970. Factors Governing the Hunting Behaviour and Selection of Food by the Great Tit (*Parus Major* L.). *Journal of Animal Ecology* 39:619–59.

Rudebeck, G. 1950. The Choice of Prey and Modes of Hunting of Predatory Birds with Special Reference to Their Selective Effect. *Oikos* 2:65–88.

Salomonsen, F. 1950. *The Birds of Greenland*. Ejnar Munksgaard: Kobenhaun.

Springer, A. M., D. G. Roseneau, E. C. Murphy, and M. I. Springer. 1984. Environmental Controls of Marine Food Webs: Food Habits of Seabirds in the Eastern Chukchi Sea. *Can. J. Fish. Aquat. Sci.* 41:1202–15.

Tinbergen, L. 1946. The Sparrow-Hawk (*Accipiter Nisus* L.) as a Predator of Passerine Birds (English summary). *Ardea* 34:184–209.

———. 1960. The Natural Controls of Insects in Pine Woods. I. Factors Influencing the Intensity of Predation by Songbirds. *Arch. Neerl. Zool.* 13:265–336.

Tinbergen, N. 1951. *The Study of Instinct*. Oxford: Oxford University Press.

———. 1953. *The Herring Gull's World*. New York: Basic Books.

Tinbergen, N., M. Imprekoven, and D. Franck. 1967. An Experiment in Spacing Out as a Defense against Predation. *Behavior* 28:307–21.

12 Human Ecology and Conservation

[Editor's note: Portions of this chapter were previously printed in Rhodora *(Drury 1980) and in an Island Institute publication (Drury and Wayne 1984). Permission to reprint the material in this form is acknowledged and appreciated.]*

An important portion of the order ascribed to landscapes is supplied by the perceptions of the human observer. Keystone species in vegetation made up of relatively few species attract attention and are called dominant or primary. These have been called successful, as if dominance and extensive geographic range were products of evolutionary success. Their broad extent can be seen as coincidental to the extent of suitable habitat available. Some species are called rare, yet most species occur in relatively small numbers. And, for our own reasons, we call some species attractive and others weeds or pests.

Many species have become conspicuous and influential because their activities impose order on their surroundings. It is important to realize that conspicuousness constitutes neither success nor failure. To suggest that it does would be to imply belief in directedness in organic processes, a notion that virtually all current evolutionists reject out of hand. Humans are one example of a general phenomenon of being numerous, conspicuous, and influential, but we are unique because we have the ability to form hypotheses and deduce conclusions from them. This ability has led some humans to have a sense of responsibility for less fortunate members of our own and other species.

Humans can occupy almost any habitat and move freely among different habitats. We carry adaptation even further, altering habitats for our own advantage. Humans have, moreover, made an additional major leap. We do not need to wait for ultraslow, undirected variation and Darwinian selection. Humans can identify problems and develop solutions to them and then pass the solutions on to friends and relations. Among human cultures, the inheritance of acquired characteristics, or Lamarckian evolution, is now a reality, greatly increasing the rate of change. We are not a specially created miracle, nor are we an aberration, but we are very, very special.

THE ECOLOGICAL IMPERATIVE FOR TODAY'S CONSERVATION MOVEMENT

The philosophical basis of much of what passes as the conservation movement lies in classical sociopolitical attitudes. The ideas that the founders of the movement fixed on were a sample of those prevalent at the end of the nineteenth century, because that is when conservation was first widely intellectualized. Similar attitudes were confirmed in the 1960s, when the movement had its renaissance. As I discussed earlier, the climax and its equilibrium conditions seem to be direct descendants of the Divine Order or Nature's Plan, which Western people inherited from their prescientific past. Succession, which acted on the community level, in contrast to selection, which operates on the individual level, replaced the Divine Hand as the mechanism to execute some sort of general plan.

The precepts of the environmental movement of the last thirty years have referred to "holistic" characteristics of stable communities. The most frequently quoted ecologists of the 1960s endorsed these ideas at the time when the philosophical "bud" of the ecological movement was unfolding. There were vigorous voices in opposition available, but few listened.

The model of stable communities, which has been largely discredited during the last two decades, is still a central element of the rhetoric of conservationists. One hears repeatedly that each species is a critical element of its community, and that if we do not know what its function is, we must not tamper for fear that we may do irreparable damage. It is stressed that each species fills a function in a holistically organized community whose sum creates a circumstance that guarantees the needs of all component parts. If one pulls at one part, one affects all other parts. This ecological imperative was translated into a policy of preservation: leave nature to her own devices, let her heal the human-inflicted wounds in her own way.

The ecological imperative depends upon the early-twentieth-century ecological models that used the theory of succession to argue that with passage of time, communities achieve a particular, preferred configuration of species characterized by a variety of "good things": maximum productivity, diversity, efficiency, large biomass, nutrient cycling, stability, "information content," and so on. These are the climaxes. Disturbance of a climax was damage, which, it was believed, set in motion forces that led to recovery. In the same model it was believed that reduction of or excess of populations initiated compensatory mechanisms and led to reestablishment of equilibriums. It has been asserted that before the advent of European civilization natural populations were saturated and stable at the carrying capacity of the environment.

Because order was believed to be established in natural communities through natural developmental processes, conservation is based on protecting natural systems from disturbing influences and restoring natural balances. Species diversity is believed to be inherently important for healthy community functioning, hence it is important to design preserves that avoid loss of species.

In direct contrast to this deterministic view, experience indicates that one can seldom prepare a model of a system a priori that will predict the effects of manipulation of parts of a natural system. An experienced naturalist's intuition, however, will do just that. What effects will be caused by removal of a species? Will the extinction of a conspicuous species have any effect?—for example, disappearance of American chestnut from the Appalachian forests or of herring gulls and great black-backed gulls from the New England shore in the late nineteenth and early twentieth century?

Traditionally, ecological research has been carried out independently of studies on the effects of human occupation. In most cases ecologists have gone out of their way to avoid having their study areas show any influence of human activity, because such places would not be "natural." In fact Paul Errington, an important figure in the history of ecological theory and game management, wrote of the pricelessness of unaltered nature because he considered such places to be the only proper natural laboratory in which a scientist could investigate the true circumstances under which natural systems have developed. Inevitably in this scheme, humans are considered to be unnatural and unwelcome elements. Accordingly, human-influenced systems must be studied separately.

Much of the activity in the conservation movement in the first years related to protection: of birds from year-round hunting; of future national parks from invasion by developers; of wetlands from drainage by farmers or from developers of shopping centers. In the 1950s the focus shifted to protection of suburban open spaces from urban sprawl, the reduction of air pollution, and protection of the population at large from the fallout of persistent pesticides and nuclear wastes.

Obstructionism is an effective program if one is small, but unfortunately the environmental establishment has persisted in this strategy even as the movement has grown beyond its limited beginnings. Much emotional energy has gone into the protection of perceived "wilderness," into crusades against the use of all poisons, and more recently into an obsession with "endangered species." Each of these concepts is all-too-often treated as an absolute, and virtually all the actions suggested around them still constitute laissez-faire protectionism.

A conservation system based on ecological imperative—ideas derived from outdated ecological models—will eventually backfire. Conservationists must seek a more solid foundation on which to build our case.

AN ALTERNATIVE APPROACH:
PROBABILISTIC CONSERVATION

The "ploy" involved in the succession-climax constellation of ideas has certainly served useful purposes in the politics of resource management, but the theory limits the sorts of questions one can ask about why and how species interact. The simplistic idea that evolutionary development of community adaptations is largely or entirely responsible for the complex structure and efficient operation of "climax" communities has discouraged pursuit of questions related to how processes and influences outside of vegetation affect species richness and productivity.

I favor an intellectual model based on assumptions consistent with the Darwinian theory of descent with gradual modification under the influence of natural selection. In this model there is no reason why humans should be excluded from a study of natural events. The effects of human activities should be recognized as providing opportunities to get valuable observations of how systems naturally subject to modification and change work. Human effects may alter, but they do not necessarily harm, a natural system, nor do they necessarily change the system more than do other primary or keystone species.

My model is one of the ecology of open systems, systems that change in space and time. Among the assumptions of this model is the idea that, because natural systems encompass a large amount of variation in their form and constituent species, successful individuals will probably use a variety of resources spread over some geographic range. Redundancy of units of habitat seems to be important for the survival of populations, as any local population may go extinct and be replaced by individuals in neighboring populations that temporarily have a surplus. The ways in which individuals move over a geographic area in space and time show how they crop several systems so as to profit by the good times in some and avoid being caught in the bad times of others.

Evidence collected by climatologists, oceanographers, and geomorphologists supports the belief that most natural processes are constantly in flux (Drury 1980). Other scientists have explicitly stated that no natural systems are closed (von Bertalanffy 1950, Chorley 1962, Drury and Nisbet

1972). This becomes clear if one reasons that the survival of all individuals is at least partly dependent upon the availability of food, the climate, and the presence or absence of other individuals, both friend and foe. Thus, if there is an environmental change, we should expect some change in population number; and since the environment is demonstrably dynamic, population levels and species compositions will constantly fluctuate.

PROBABILISTIC SOLUTIONS TO THE PROBLEMS OF DETERMINISTIC CONSERVATION

Although deterministic models and theories have continually been challenged and refuted, phrases such as "climax state," "stability of ecosystems," "carrying capacity," and "balance of nature" still form the core of much conservation and environmental rhetoric. This has two primary unfortunate consequences.

Intervention versus Preservation

The first problem results from the application of deterministic models to nature conservation and land use. If, as deterministic models suggest, there is a natural balance or equilibrium to which tracts of land or ecosystems naturally tend, then a laissez-faire conservation policy, or "preservation," would be the best practice. Any human-related activities would disrupt the "natural balance."

This hesitance to interfere or actively manage nature preserves by manipulating populations or habitats has led to problems for preservationist organizations and government agencies. If such an organization or agency buys a tract of land to save or protect a population of rare or endangered species, we cannot be certain that a laissez-faire policy will in fact preserve or protect that species from extinction.

Control of gull populations on Petit Manan Island, Maine, provides an excellent example in which knowledge of species' ecology combined with a willingness to actively intervene allowed the successful recolonization of vulnerable species, terns in particular, to a traditional nesting colony.

Part of the story of the decline and subsequent increase of herring gull populations in New England has been discussed in chapter 9 (see also Drury 1973, 1974). In 1968 approximately 1,400 pairs of terns had nested on Petit Manan, while about 150 pairs of herring and great black-backed gulls nested on neighboring Green Island (Hatch 1970). Human occupancy of Petit Manan ended in 1974 after the island's lighthouse was automated. Gull numbers increased to about 350 pairs on Green Island and 10 pairs had settled on Petit Manan in 1977 (Korschgen 1979). By 1978 terns and laughing gulls began to decrease and by 1983 the terns had abandoned the island.

In 1984 some of my students looked into what was happening to the displaced Petit Manan terns, some of which settled on Flat and Nash Islands about ten miles to the east, while others settled on Egg Rock in Frenchman Bay about twelve miles to the west. The students camping on Flat and Nash Islands reported that herring gulls, after six days of fog had denied them access to their usual food from lobster boats, began to hunt in the tern nesting area and eat all the young chicks. The students at Egg Rock reported herring gulls pouncing on incubating adult laughing gulls and terns from the air and eating them alive.

The temporary increase in numbers of terns nesting on many islands made it difficult to convince people that the tern population was in trouble. From some peoples' limited view, the terns were in good shape. This point emphasizes the need to keep track of terns over a long stretch of coast, and to realize that subpopulations are strongly influenced by immigration and emigration.

Gull population studies conducted in the 1960s and 1970s indicated that a relatively small number of gulleries produce 85 percent of the young birds, and that this keeps the population going (Kadlec and Drury 1968; Drury and Nisbet 1972). Reproduction on other islands is not important. This means two things, first that eliminating gulls from a number of outer islands where terns might nest will have no effect on "reducing the gull population." It also suggests that our concern for terns must be focused on colonies on those islands that are successful. Those islands where parents produce excess young are significant in maintaining tern populations, but keeping terns nesting on islands where parents fail

Plate 6. The William H. Drury Jr. Biological Research Station on Petit Manan National Wildlife Refuge. Photo courtesy of John Anderson.

provides a population sink. This seems to be the fate of the subpopulations that "bivouac" for a few years, scattered on many islands after being driven off a "populous center," as was the case with the islands studied by my students.

In mid-May 1984, the U.S. Fish and Wildlife Service eliminated herring gulls from Petit Manan and Green Island using the toxicant DRC-1339, which had been shown from earlier gull control studies to be effective and humane and to only affect target species. By early June over 500 terns and 60 laughing gulls were seen taking up nesting territories on Petit Manan. By the end of June, crews estimated 855 tern nests and 200 laughing gull nests on Petit Manan. In July, the "bivouac" sites on Nash and Flat Islands broke up and a corresponding number of nesting pairs settled in on Petit Manan. The ternery on Petit Manan was restored.

[Editor's update: Researchers have returned to the island each summer since the initiation of the project, and in 1995 there were over 2,000 nesting pairs of terns. In addition, increasing numbers of Atlantic puffins, Leach's storm petrels,

laughing gulls, and black guillemots have occupied the island. Other managed islands in the Gulf of Maine have had similar success, while populations at unmanaged sites have dwindled and vanished.]

Some argued against killing the gulls on humanitarian grounds. I agree: killing is not a pretty or enjoyable activity. But gulls eating living tern chicks is not a pretty sight, nor is the sight of a herring gull pecking at the bleeding head of a living laughing gull that was just recently incubating its eggs.

I think that the philosophical question of killing one species to favor another was made and accepted by those early agriculturalists who pulled up plants that inhibited the growth of their crops—they weeded the garden. A biologist can argue that it is precious and self-serving to make a philosophical separation of plants from animals, or "lower animals" from vertebrates, or us. Why should fish be placed "beyond the pale?" When we find vulnerable species that we think are important and we want to encourage them, we may have to weed the garden.

Some people hold nature to be in an almost hallowed state, while imposition of an order that favors humans desecrates that natural condition. Some would sacrifice the welfare of vulnerable species to maintain the purity of their ideals. These ideals, as constructs of the human imagination, are easily insulated from testing against reality.

Under present environmental conditions, laissez-faire protection without intervention may not work to protect desired species. The differences between Darwinian and non-Darwinian approaches to conservation were nowhere more clearly expressed than in attitudes toward rehabilitation of peregrine falcons and California condors in the wild. The issue came into sharp focus when Tom Cade of Cornell University, representing a group of falconers, came to Massachusetts Audubon seeking support for a project to breed peregrines in captivity and release the offspring into the wild. Tom's idea seemed to be a fine one to me and to others at the Massachusetts and National Audubon Societies. Other conservationists, however, by no means agreed.

The peregrine population was much reduced all across North America and Europe, and what breeding populations remained were widely separated from each other. Thus they were subject to what I consider the most important biological issue involved in really rare species: reduced local genetic variability because of the smallness of the population and low probability of exchange of genetic material. Those concerned with the welfare of the eastern peregrine argued that genetic variability should be deliberately promoted in the breeding stock (Clement 1974). This could be promoted by using captive breeders taken from the wild in several parts of the world—crossing birds from Spain with ones from Norway, Arctic Canada, and Idaho, for example. Turnover in the captive stock should be encouraged by introducing new breeders from the wild. The goal should be releasing as many offspring as possible into the wild and then letting natural selection take its course.

An important element of this program was the acceptance of the possibility of a high initial mortality of young birds. Many individuals would prove to be unfit for life in the modern, partially urbanized East Coast. If enough of a genetic mix was available to begin with, however, we had hopes that a locally adapted population would eventually emerge.

Efforts to breed peregrines in captivity proved successful. By the summer of 1991, five traditional nesting sites had been reoccupied in Maine, and dozens more were established along the eastern seaboard of the United States. In contrast, preservationists actively opposed captive breeding of California condors. Condor populations continued to decline until at last even the most strenuous opponents of captive breeding were persuaded and the remaining condors were captured and placed in breeding facilities. The outcome of this program remains uncertain at this time.

[Editor's update: Since 1991 peregrine numbers in the lower forty-eight states have continued to increase. In 1995 there were thirty-three territorial pairs in New England alone. The U.S. Fish and Wildlife Service has filed an advance notice of intent to delist the species from its endangered status (pers. comm. US-FWS and 34406 Fed. Reg., June 30, 1995).

As a compromise with noninterventionists, biologists were finally able to collect California condor eggs from the wild for use in a captive breeding program, but the wild population continued to decline drastically. In 1987 the last of the wild condors were captured, resulting in a total captive population of twenty-seven birds. As of April 1996 there were eighty-six condors living in captivity. Release of birds began in 1992 and as of 1996 there were seventeen captive-hatched con-

dors living in the wild. Before release the birds undergo aversion training to teach them to avoid humans and power lines. Supplemental feeding by the U.S. Fish and Wildlife Service is the primary food source for the wild birds and serves to reduce the risks of lead poisoning (USFWS 1996; see also Snyder and Snyder 1989).]

It is interesting and revealing to note that on many preserves where active management has been practiced (such as captive breeding of the peregrine falcon and the culling of gulls on Petit Manan), populations have been saved and revived. Ecologically speaking, these activities created environments to which certain species were adapted. By understanding the requirements of different species and the effects different activities have on them, ecologists were able to promote desired species. In a sense, this is the aim of probabilistically based ecology and conservation. By acquiring information from careful observation and experimentation, ecologists and others involved in conservation can begin to cooperate with nature, carefully guiding it as stewards, so that it may best meet humanity's needs and values.

Values are the key to understanding a conservation system based on probabilistic models. Probabilistic ecology does not suggest an ecological imperative. There are no balances to protect. Rather, nature is protected and promoted because we derive benefits from it, whether the benefit be aesthetic, spiritual, scientific, or economic.

In trying to justify nature conservation, it is often argued that species need protection in order to maintain ecological diversity. The rationale for diversity has two roots. The first is derived from an old, deterministic model that suggested diversity leads to stability. This has been disproven. Snail darters need not be protected to maintain the stability of ecosystems. The other argument for diversity stems from the claim that we must maintain a large gene pool. The larger our gene pool, the greater the potential for finding new food, medicine, and energy sources. Although this is a valid argument, it is not an ecological imperative. Rather it is just another human-based value. As Rene Dubos wrote, "Conservation is based on human value systems; its deepest significance is in the human situation and the human heart. Saving marshlands and redwoods does not need biological justification any more than does opposing callousness and vandalism" (1972).

Humans as Part of Nature

The second problem with deterministic ecology and conservation has already been alluded to. Notions of balance tend to separate humans from nature. This attitude implies that humans and their habitats are not natural because they have consistently created imbalances. Environmentalists continually assert that humans and their technological society have destroyed nature. They argue that before humans appeared on the scene, all was peaceful and harmonious. This is a distorted, unsubstantiated view. I believe that the dichotomy between human-influenced systems and "natural" systems is not realistic or helpful and that it leads to an unjustified pessimism among environmentalists.

The separation of humans from nature has many additional implications. Not only does it prevent us from achieving our desired goals in conservation, as discussed above, but on a more philosophical level, it relieves us of any responsibility for nature. We are not part of nature; we just use it, or "protect" it by keeping people apart from nature preserves.

This attitude is indirectly perpetuated by our federal lands policies. The Wilderness Act, for example, defines wilderness areas as places where "the earth and its community of life are untrammeled by man, where man himself is a visitor who does not remain" (16 U.S.C. Section 1131[a]). This must leave visitors with the feeling that humans and nature cannot coexist.

If a preservationist approach tends to separate humans from nature, then the opposite may also be possible. If humans and their activities were integrated into some park system, visitors might leave with new ideas regarding nature and their place in it. This seems to be the trend of some European conservation systems.

HUMAN ECOLOGY AND SOCIETY'S TOOLS

After extensive study of natural selection, most students doubt (as did Darwin) that natural selection can select one species for the benefit of another species. The Welsh botanist John Harper has cogently stated these opinions: "A theory of natural selection that is based on the fitness of in-

dividuals leaves little room for the evolution of populations or species toward some optimum, such as better use of environmental resources,
higher productivity per area of land, more stable ecosystems, or even for
the view that plants in some way become more efficient than their ancestors. Instead, both the study of evolutionary processes and of the natural
behaviour of populations suggest that the principles of 'beggar my neighbor' and 'I'm all right Jack' dominate all and every aspect of evolution"
(1977). Many conservationists plead: "let nature take her course"; an approach reminiscent of proponents of the free market who say: "let the
market act."

The market works, no doubt about that, but it works in the ways that
natural selection works. As with natural selection, pressures to meet short-
term goals inhibit or prevent attainment of long-term optimization when
the improvement requires a system to pass through temporarily non-
adaptive conditions. The collapse of efficient public transportation in the
face of the selfish convenience of automobiles and the manipulation of
self-serving big businesses illustrates the actions of the market in terms
of the long range "public good."

These pressures act on conservationists as well. The press of economic
necessity pushes most conservation organizations to concentrate on local
concerns, because volunteers prefer to work in their own direct interests.
The need for financial support stimulates them to modify their philoso-
phies: "We are interested in people, not just birds." Over time, some organizations have grown at the expense of others; the ones that have grown
are primarily those that are in effect insurance agencies for the environmental amenities of affluent suburbanites. It becomes clear that raising
funds and becoming successful in legislative and legal arenas depends
upon assertiveness and confidence in one's cause. There is little open
questioning of philosophical or scientific justification despite the scientific
revolutions that are going on.

Our legislative and legal systems supply rules for the arena where selfish interests compete. No single special interest will yield its agenda to the
general good. For example, the discussions that led to the banning of
pesticides showed that legislators would require chemicals be available
to all of the public or to none, because that is what the special interest

groups demanded. The idea of restricted use in case of serious need was repugnant to those who did not trust the decision makers. Each group insisted on defining "real need" themselves. So individuals in politics tend to manipulate legislative acts in their own interests, and it is the function of the courts to interpret those acts.

I see society as made up of networks of interactive subsystems—legal, industrial, legislative—more or less coupled together. Some of the subsystems operate largely in isolation; others have wide influence. Most of our subsystems (social institutions) concentrate on serving their members. Most operate under their own dogmatic rules. Within each social institution there tends to be an establishment that is defensive about changing the rules for fear that its influence will be diminished. So the operators of our institutions try to suppress debate on the bases of their actions. Industry has its proprietary information; the governments have their classified material; the academics have their authorities. What chance is there to reexamine assumptions and data? The legal imperative, the legislative imperative, and the market imperative, like ecological adaptations, are patched onto existing structures and serve selfish interests. Our rules were patched on as our society outgrew the villages where everyone knew each other, knew everyone's past behavior, and could bring social pressure on antisocial individuals to conform to group norms. Maybe our most serious problem results from the opportunities offered by anonymity.

Our social institutions originated as mechanisms to gain selfish ends. They will not solve our problems any more than natural selection will. We need another avenue of approach, a new perspective and the will to direct our institutions to performing socially creative functions.

A NEW APPROACH

A current that flows counter to humans' selfishness has run through the centuries despite our legislatures, courts, and markets. This current is the main characteristic that makes humans special and gives us hope and optimism. The current is *a sense of responsibility based on the ability to form hy-*

potheses. It can find its biological roots in the actions of leaders who undertook some degree of "reciprocal altruism" for the larger group. This worked because the group gave loyalty and gratitude in return for care. In small in-breeding groups a sense of responsibility among the powerful on the one hand and of loyalty among the followers on the other obviously increased "fitness." This responsibility may have stretched beyond the extended family group when it was advantageous for several groups to join in hunting parties for big game in the Pleistocene. The story of human progress since then can, with some justification, be presented as an expansion of the sense of responsibility.

A major step in the neolithic revolution was the domestication of animals and cultivation of plants. This tremendous step required assumption of responsibility for organisms that came to depend upon us and that are not members of our species. One must protect the livestock from depredations of wild animals and protect crops from interference by weeds. I doubt that, in those dim, long-lost days, philosophical questions were raised about whether it was right to favor some species over another or to interfere with the "natural balance."

This period of transition from slave to master of the environment added an important dimension to responsibility as larger agrarian communities grew. Power accreted to certain individuals, and with the increasing size of groups the concept of "we" and "they" (simple in tribal units) became complex, especially as trade made it of interest to maintain associations with the people with whom we trade. Another major revolution in philosophy is associated with the question "And who is my brother?" As the sense of responsibility has expanded to protect even those whom we considered to be our enemies, we subvert the very foundation of natural selection. That is real progress.

The Darwinian revolution challenged the easy confidence people had that humans are a special creation and, that ultimate arrogance, "created in God's image." Now, if we are descended from apes, the question of who is my brother and who is my neighbor becomes even more awkward. The social revolution that I see in the environmental movement gropes toward "internalizing" the implications of evolution in social thought. Established religions have bitterly resented this. Where do our responsibilities stop? If in our garden we feel free to pull up a wildflower that we call a weed, do we feel guilty killing a herring gull that would otherwise drive Arctic terns and laughing gulls from their nesting grounds? Do apes have special value? Do mammals? Do vertebrates? Do animals over plants?

We are committed to a debate among scientists, among conservationists, and among members of the public as to what we think is right for humans in their interactions with their habitat. Some would have native people kept as relics of traditional ways of life. Similarly, some consider their responsibilities done when they have established wilderness where the affluent can enjoy their safaris or canoe trips. Are those who rest after setting aside sanctuaries in suburbia not saying, "I'm all right, Jack"?

Things are happening quickly and ideas are changing quickly. Many thoughtful people are groping for whatever social institutions are available —political, legal, economic, intellectual—to ensure that values not yet codified nor recognized are not destroyed in the rush of indifference. These people are in a true Darwinian sense using what they have inherited to measure their fitness (the survival of their ideas) against the background of their habitat (the social realities) in which they find themselves. They use whatever tools are available, and we should wish them Godspeed as a holding action, even though many of the biological ideas now being used are as archaic as the social institutions and constraints within which they act.

It seems clear that the ecological imperative will not long be viable scientifically. We are disillusioned with our social institutions and imperatives such as the legal arena, the marketplace, and the legislatures. Yet those institutions are the stage upon which we must act. They are the stuff upon which social pressure acts. We now know that many people, surprisingly many people, care for certain kinds of interaction with nature. Thoreau and Leopold took steps to codify some of these ideas as a conservation ethic. This sympathy based on intuition is an imperative separate from the market, the legal, or the ecological. It constitutes a sense of responsibility both to nature and to human interests. Why should we accept it as "given" that this responsibility is secondary to the market as expressed by social rules that favor those who would use land for profit as opposed to those who would husband the land?

Most of our paradigms do not tolerate dissidents. They urge us to let nature or industry follow the rules that they know will work—to "cop out." The problems are ethical and pragmatic. I think we should remember the precept of those who design women's fashions. It takes love and perspicacity, but we believe that once we know what we are doing, nature can be improved upon.

CONCLUSION

These are very active decades in conservation. Decisions are being made that affect how parcels of land will be treated. While I do not take the "doomsayer's" attitude that what is lost now is lost forever (that is, a Clementsian, deterministic view), I am interested in the consequences of actions and in learning to predict what may happen in the future by observing the effects of certain sorts of actions. Through this process I hope to be able to suggest some principles of sympathetic management.

If the coming years are to be ones of enlightened management of the land, then that management should be based on developing understanding of species behavior. It is therefore important to investigate the consequences of several different kinds of human activities. If we can predict the consequences of our actions, we can make enlightened judgments instead of leaning on the mysticism of "nature's way." Obstructionism or preservationism may be effective for a small budget and an unenlightened public during the first decades of an environmental movement, but ecological laissez-faire will not do as our sophistication grows. Once we have accepted a new paradigm based on "humanized" scientific understanding of our land and the other species we live with, we must use this new understanding to foster new solutions. As to the philosophical issues of nature's order and "playing God," we now know that laissez-faire ecology, like laissez-faire economics, doesn't lead to balanced systems, it leads to monopolies. Unless we believe that there is a natural order established at the Creation, we should acknowledge that when we won't play God, someone else will.

. . .

To sum up: chance and change are ubiquitous; habitats are heterogeneous; selection drives parents to produce a great excess of young; death (disturbance) is necessary for life; and movements of individuals are pervasive. That's the way the world is made and works. I think, in fact, that these, not order and integration, are what allow species to ride out the tribulations of this imperfect world.

SOURCES

Chorley, R. J. 1962. *Geomorphology and General Systems Theory.* U.S. Geol. Survey Prof. Paper 500-B. Washington, D.C.: USGPO.
Clement, R. C. (ed.) 1974. *Proceedings of a Conference on Peregrine Falcon Recovery, 13–15 February 1974.* Greenwich, Conn.: National Audubon Society.
Drury, W. H. 1973. Population Changes in Seabirds in New England. Part 1. *Bird-Banding* 44 (4): 267–313.
————. 1974. Population Changes in Seabirds in New England. Part 2. *Bird-Banding* 45 (1): 1–15.
————. 1980. Rare Species of Plants. *Rhodora* 82:3–48.
Drury, W. H., and I. C. T. Nisbet. 1972. The Importance of Movements in the Biology of Herring Gulls in New England. In *Population Ecology of Migratory Birds: A Symposium,* 173–212. U.S. Department of the Interior. Wildlife Research Report 2.
Drury, W. H., and P. Wayne. 1984. Theoretical Basis for Maine Island Research. In *People and Islands,* edited by P. W. Conkling, R. E. Leonard, and W. H. Drury. Rockland, Maine: Island Institute.
Dubos, R. 1972. *A God Within.* New York: Charles Scribner's Sons.
Errington, P. 1957. *Of Men and Marshes.* New York: Macmillan.
Harper, J. 1977. *Population Biology of Plants.* London: Academic Press.
Hatch, J. J. 1970. Predation and Piracy by Gulls at a Ternery in Maine. *Auk* 87:244–54.
Kadlec, J. A., and W. H. Drury. 1968. Structure of the New England Herring Gull Population. *Ecology* 49 (4): 644–76.

Kadlec, J. A., W. H. Drury, and D. K. Onion. 1969. Growth and Mortality of Herring Gull Chicks. *Bird-Banding* 40 (3): 222–33.

Korschgen, C. E. 1979. *Coastal Waterbird Colonies: Maine.* U.S. Fish and Wildlife Service, Biol. Serv. Program. FWS/OBS-79/09.

Snyder, N. F. R., and H. A. Snyder. 1989. Biology and Conservation of the California Condor. *Current Ornithology* 6:175–267.

U.S. Fish and Wildlife Service. 1996. *California Condor Recovery Plan.* 3rd rev. Portland, Ore.

U.S. Fish and Wildlife Service. July 1996. Personal communication. Michael Amaral, New England Field Office.

von Bertalanffy, L. 1950. The Theory of Open Systems in Physics and Biology. *Science* 111:23–29.

When your views on the world and your intellect are being challenged and you begin to feel uncomfortable because of a contradiction you've detected that is threatening your current model of the world or some aspect of it, pay attention. You are about to learn something.

WILLIAM H. DRURY JR.

Appendix

INVERTEBRATES

Snails, Class Gastropoda

periwinkle	*Littorina littorea*
moon snail	*Polinices* spp.
dog whelk	*Thais lapillus*
channeled whelk	*Busycon canaliculatum*
marsh snail	*Melampus bidentatus*

Bivalves, Class Bivalvia

blue mussel	*Mytilus edulis*
marsh mussel	*Modiolus demissus*
bay scallop	*Aequipecten irradians*
Greenland cockle	*Serripes groenlandicus*
quahog	*Mercenaria mercenaria*
arctic species of clam	*Mesodesma arctatum*

Worms, Class Polychaeta

bloodworm	*Glycera dibranchiata*
sandworm	*Sabellaria vulgaris*
tube worm	*Spirorbis* & related genera

Horseshoe Crabs, Class Merostomata

horseshoe crab	*Limulus polyphemus*

Crustaceans, Class Crustacea

barnacle	*Balanus balanoides*
sand flea	Order Amphipoda
grass shrimp	*Palaemonetes pugio*
spider crab	*Libinia emarginata*
fiddler crab	*Uca pugnax*
northern lobster	*Homarus americanus*

Sea Stars, Class Stelleroidea

starfish	Subclass Asteroidea

Sea Urchins, Class Echinoidea

purple sea urchin	*Arbacia punctulata*
green sea urchin	*Strongylocentrotus droebachiensis*

Insects, Class Insecta

monarch butterfly	*Danaus plexippus*
yellow jackets	*Vespula maculifrons*
water flea	*Daphnia* spp.
spruce budworm	*Choristoneura fumiferana*

FISH

Eels, Order Anguilliformes

American eel	*Anguilla rostrata*

Herring Family—Clupeidae

alewife	*Alosa pseudoharengus*
American shad	*Alosa sapidissima*
Atlantic menhaden	*Brevoortia tyrannus*
Atlantic herring	*Clupea harengus*

Salmon Family—Salmonidae

Atlantic salmon   *Salmo salar*

Smelt Family—Osmeridae

rainbow smelt   *Osmerus mordax*

Cod Family—Gadidae

Atlantic cod   *Gadus morhua*

Killifish Family—Cyprinodontidae

sheepshead minnow   *Cyprinodon variegatus*
mummichog   *Fundulus heteroclitus*

Silverside Family—Atherinidae

Atlantic silverside   *Menidia menidia*

Bass Family—Percichthyidae

striped or "sea" bass   *Morone saxatilis*

Bluefish Family—Pomatomidae

bluefish   *Pomatomus saltatrix*

Right-Eye Flounder Family—Pleuronectidae

winter flounder   *Pseudopleuronecetes americanus*

Darter and Perch Family—Percidae

snail darter   *Percina tanasi*

B I R D S

Auk Family—Alcidae

black guillemot	*Cepphus grylle*
Atlantic puffin	*Fratercula arctica*

Cormorant Family—Phalacrocoracidae

great cormorant	*Phalacrocorax carbo*
double-crested cormorant	*Phalacrocorax auritus*

Swan, Goose, and Duck Family—Anatidae

white-winged scoter	*Melanitta deglandi*
surf scoter	*Melanitta perspicillata*
black scoter	*Melanitta nigra*
old squaw	*Clangula hyemalis*
common eider	*Somateria mollissima*
king eider	*Somateria spectabilis*
red-breasted merganser	*Mergus serrator*
Canada goose	*Branta canadensis*

Shearwater Family—Procellariidae

shearwaters	*Puffinus* spp.

Petrel Family—Hydrobatidae

Leach's storm petrel	*Oceanodroma leucorhoa*

Albatross Family—Diomedeidae

albatross	*Diomedea* spp.

Gannet Family—Sulidae

northern gannet	*Morus bassanus*

Jaeger Family — Stercorariidae

pomarine jaeger	*Stercorarius pomarinus*

Gull and Tern Family — Laridae

great black-backed gull	*Larus marinus*
herring gull	*Larus argentatus*
laughing gull	*Larus atricilla*
black-legged kittiwake	*Rissa tridactyla*
common tern	*Sterna hirundo*
arctic tern	*Sterna paradisaea*
roseate tern	*Sterna dougallii*
sooty tern	*Sterna fuscata*
bridled tern	*Sterna anaethetus*

Heron Family — Ardeidae

green heron	*Butorides striatus*
black-crowned night heron	*Nycticorax nycticorax*
snowy egret	*Egretta thula*
great blue heron	*Ardea herodias*

Plover Family — Charadriidae

gray plover	*Pluvialis squatorola*
golden plover	*Pluvialis dominica*
killdeer	*Charadrius vociferus*

Sandpiper and Phalarope Family — Scolopacidae

Baird's sandpiper	*Calidris bairdii*
semi-palmated sandpiper	*Calidris pusilla*
white-rumped sandpiper	*Calidris fuscicollis*
sanderling	*Calidris alba*
red phalarope	*Phalaropus fulicarius*
northern phalarope	*Lobipes lobatus*
Wilson's phalarope	*Steganopus tricolor*
lesser yellow legs	*Tringa flavipes*

Hawk and Eagle Family—Accipitridae

Cooper's hawk   *Accipiter cooperii*
northern goshawk   *Accipiter gentilis*
red-shouldered hawk   *Buteo lineatus*
bald eagle   *Haliaeetus leucocephalus*
osprey   *Pandion haliaetus*

American Vulture Family—Cathartidae

turkey vulture   *Cathartes aura*
California condor   *Gymnogyps californianus*

Falcon Family—Falconidae

peregrine falcon   *Falco peregrinus*
sparrow hawk   *Falco sparverius*

Owl Family—Tytonidae

snowy owl   *Nyctea scandiaca*

Pigeon Family—Columbidae

pigeon   *Columba livia*

Lark Family—Alaudidae

horned lark   *Eremophila alpestris*

Swallow Family—Hirundinidae

tree swallow   *Iridoprocne bicolor*

Swift Family—Apodidae

common swift   *Apus apus*

Crow and Jay Family—Corvidae

American crow	*Corvus brachyrhynchos*
blue jay	*Cyanocitta cristata*

Titmice Family—Paridae

black-capped chickadee	*Parus atricapillus*
tufted titmouse	*Parus bicolor*
great tit	*Parus major*
coal tit	*Parus ater*
blue tit	*Parus caeruleus*

Wren Family—Troglodytidae

Carolina wren	*Thryophorus ludovicianus*
marsh wren	*Cistothorus palustris*

Mockingbird Family—Mimidae

northern mockingbird	*Mimus polyglottus*

Thrush Family—Turdidae

European robin	*Erithacus rubecula*
hermit thrush	*Catharus guttatus*

Wood Warbler Family—Parulidae

wood warbler	*Phylloscopus sibillatrix*
yellow-rumped warbler	*Dendroica coronata*
black-throated green warbler	*Dendroica virens*
blackburnian warbler	*Dendroica fusca*
northern parula warbler	*Parula americana*
northern waterthrush	*Seiurus noveboracensis*
American redstart	*Setophaga ruticilla*

Starling Family—Sturnidae

European starling   *Sturnus vulgaris*

Weaver Finch Family—Ploceidae

house sparrow   *Passer domesticus*

Sparrow, Finch, And Bunting Family—Fringilllidae

Lapland longspur   *Calcarius lapponicus*
dark-eyed junco   *Junco hyemalis*
cardinal   *Richmondena cardinalis*
white-winged crossbill   *Loxia leucoptera*
evening grosbeak   *Coccothraustes vespertinus*
white-throated sparrow   *Zonotrichia albicollis*
tree sparrow   *Spizella arborea*
chaffinch   *Fringilla coelebs*
yellow bunting   *Emberiza citrinella*
rufous-sided towhee   *Pipilo erythrophthalmus*

Grouse Family—Phasianidae

ring-necked pheasant   *Phasianus colchicus*
ruffed grouse   *Bonasa umbellus*

M A M M A L S

Dog Family—Canidae

arctic fox   *Alopex lagopus*
red fox   *Vulpes fulva*

Cat Family—Felidae

Canada lynx   *Lynx canadensis*

Earless Seal Family—Phocidae

harbor seal   *Phoca vitulina*

Mouse Family—Muridae

brown lemming   *Lemmus trimucronatus*
voles   *Microtus* spp.
house mouse   *Mus musculus*

Rabbit Family—Leporidae

snowshoe hare   *Lepus americanus*

PLANTS

Horsetail Family—Equisetaceae

river horsetail   *Equisetum fluviatile*

Pine Family—Pinaceae

balsam fir   *Abies balsamea*
white spruce   *Picea glauca*
red spruce   *Picea rubens*
black spruce   *Picea mariana*
hemlock   *Tsuga canadensis*
tamarack   *Larix laricina*
white pine   *Pinus strobus*
Scots pine   *Pinus sylvestris*
pitch pine   *Pinus rigida*

Cypress Family—Cupressaceae

arborvitae   *Thuja occidentalis*
red cedar   *Juniperus virginiana*

Cattail Family—Typhaceae

cattail   *Typha latifolia*

Pondweed Family—Zosteraceae

widgeon grass   *Ruppia occidentalis*

Grass Family—Gramineae

spike grass   *Distichlis spicata*
reed   *Phragmites communis*
American marram grass   *Ammophila breviligulata*
European marram grass   *Ammophila arenaria*
midtide grass, thatch, or thatch grass   *Spartina alterniflora*
high-tide grass, or salt marsh hay   *Spartina patens*

Sedge Family—Cyperaceae

bulrush   *Scirpus rubotinctus*
tall beaked sedge   *Carex* spp.

Rush Family—Juncaceae

black grass   *Juncus gerardi*

Orchid Family—Orchidaceae

yellow lady's slipper   *Cypripedium parviflorum*

Willow Family—Salicaceae

willows   *Salix* spp.
aspen   *Populus* spp.
balsam poplar   *Populus balsamifera*

Wax-Myrtle Family—Myricaceae

bayberry   *Myrica pensylvanica*

Walnut Family—Juglandacea

hickories   *Carya* spp.

Birch Family—Betulaceae

mountain alder   *Alnus crispa*
black birch   *Betula lenta*
yellow birch   ·   *Betula lutea* (also: B. *alleghaniensis*)
gray birch   *Betula populifolia*
white birch or paper birch   *Betula papyrifera*

Beech Family—Fagaceae

beech   *Fagus grandifolia*
live oak   *Quercus virginiana*
red oak   *Quercus rubra*
black oak   *Quercus velutina*
white oak   *Quercus alba*
scarlet oak   *Quercus coccinea*
American chesnut   *Castanea dentata*

Elm Family—Ulmaceae

elm   *Ulmus* spp.

Goosefoot Family—Chenopodiaceae

samphire   *Salicornia europaea*

Rose Family—Rosaceae

meadowsweet   *Spiraea latifolia*

hawthorne   *Crataegus* spp.
raspberry   *Rubus idaeus*
purple-flowering raspberry   *Rubus odoratus*
seaside rose   *Rosa rugosa*
blackthorn   *Prunus spinosa*

Cashew Family—Anacardiaceae

sumac   *Rhus* spp.

Maple Family—Aceraceae

sugar maple   *Acer saccharum*
red maple   *Acer rubrum*

Rockrose Family—Cistaceae

beach heather   *Hudsonia tomentosa*

Sour Gum Family—Nyssaceae

black gum   *Nyssa sylvatica*

Heath Family—Ericaceae

large cranberry   *Vaccinium macrocarpon*
small cranberry   *Vaccinium oxycoccos*

Leadwort Family—Plumbaginaceae

sea lavender   *Limonium nashii*

Olive Family—Oleaceae

white ash   *Fraxinus americana*

Milkweed Family—Asclepiadaceae

milkweed   *Asclepias curassavica*

Composite Family—Compositae

seaside goldenrod   *Solidago sempervirens*
New York aster   *Aster novi-belgii*
whorled aster   *Aster acuminatus*
large-leafed aster   *Aster macrophyllus*
marsh elder   *Iva fruticosa*
black-eyed Susan   *Rudbeckia serotina*

Peat Moss Family—Sphagnaceae

peat moss   *Sphagnum fuscum*
peat moss   *Sphagnum rubellum*

Lichen

old man's beard   *Usnea* spp.

Algae

Irish moss   *Chondrus crispus*
kelps   *Laminaria* spp. and *Alaria* spp.
bladder wrack   *Fucus* spp.
knotted rockweed   *Ascophyllum nodosum*
sea lettuce   *Ulva lactuca*
green filamentous algae   *Enteromorpha intestinalis*

About the Author

WILLIAM HOLLAND DRURY JR.
MARCH 18, 1921–MARCH 26, 1992

William H. Drury Jr. was the fifth child, first son, of the artists William H. and Hope C. D. Drury. As the only boy in the family, he learned to entertain himself by roaming the fields, marshes, and beaches of rural Middletown, Rhode Island, and sketching the animals and landscapes he found there. He was educated at home and at a neighbor's until he entered the Saint George's School, and he later spent a critical postgraduate year at Haileybury College in England, where watching birds was a recognized and respected vocation enjoyed even by the school's headmaster.

He graduated magna cum laude in biology from Harvard College in 1942 and was elected to Phi Beta Kappa. During World War II, he served as a chief quartermaster in the U.S. Naval Reserve in the European, Atlantic, Pacific, and Asiatic theaters. After the war, he returned to Harvard University and was elected a Junior Fellow in 1949. For his Ph.D. in biology and geology, Drury explored the relationship between geology and vegetation patterns in Alaska and northwestern Canada. He taught at Harvard University as an assistant professor and lecturer in biology.

The chance to create an ornithological field research station for the Massachusetts Audubon Society lured Drury away from academia for twenty years, during which he served as Audubon's director of education, director of research, and director of scientific staff. Herring gulls, friends since his early years, were his favorite research subjects, but he also spent much energy supporting Rachel Carson's views on DDT and other pesticides. He served on the Science Advisory Committees for Presidents Kennedy and Nixon, as well as on several state boards.

Drury returned to the north in the 1970s to study seabirds in Alaska's Seward Peninsula for the National Oceanic and Atmospheric Administration's Outer Continental Shelf Environmental Assessment Program. The opportunity to work at a new, ecologically focused college on the coast of Maine returned him to teaching in 1976. With his varied interests and background, he brought a unique perspective to students of "human ecology" at the College of the Atlantic. His course offerings included landforms and vegetation, populations and species, ornithology in the field, natural history drawing, and animal behavior. His research efforts were spent on pragmatic projects such as the return of terns and other seabirds to traditional nesting islands in the Gulf of Maine, restoration of peregrine falcon populations in the east, and the gathering of baseline vegetation data on Gulf of Maine islands for long-term studies. He was working on the final drafts of this manuscript at the time of his death.

Illustration, previous page: Killdeer, pen and ink.

Index

Compositor: Integrated Composition Systems, Inc.
Text: Palatino
Display: Snell Roundhand Script and Bauer Bodoni
Printer and Binder: BookCrafters, Inc.